Understanding I.T. in Const

Other titles in the series:

Understanding Quality Assurance in Construction: a practical guide to ISO 9000
H W Chung (1999)

Understanding The Building Regulations, 2nd edition
Simon Polley (2001)

Understanding Active Noise Cancellation
Colin H Hansen (2001)

Understanding JCT Standard Building Contracts, 7th edition
David Chappell (2003)

Understanding I.T. in Construction

Ming Sun and Rob Howard

Spon Press
Taylor & Francis Group

LONDON AND NEW YORK

First published 2004
by Spon Press
11 New Fetter Lane, London EC4P 4EE

Simultaneously published in the USA and Canada
by Spon Press
29 West 35th Street, New York, NY 10001

Spon Press is an imprint of the Taylor & Francis Group

Publisher's Note
This book has been produced from camera-ready-copy
supplied by the authors
Printed and bound in Great Britain by
TJ International, Padstow, Cornwall

Every effort has been made to ensure that the advice and information in
this book is true and accurate at the time of going to press. However,
neither the publisher nor the authors can accept any legal responsibility
or liability for any errors or omissions that may be made. In the case
of drug administration, any medical procedure or the use of technical
equipment mentioned within this book, you are strongly advised
to consult the manufacturer's guidelines.

British Library Cataloguing in Publication Data
A catalogue record for this book is available from the British Library

Library of Congress Cataloging in Publication Data
A catalog record for this book has been requested

ISBN 0-415-23190-6

CONTENTS

LIST OF FIGURES   xi
LIST OF TABLES   xiii
PREFACE   xiv
ACKNOWLEDGEMENTS   xv
LIST OF ABBREVIATIONS   xvi

CHAPTER 1
Overview of IT Applications in Construction

LEARNING OBJECTIVES   1
INTRODUCTION   1
1.1 THE INFORMATION AGE   1
1.2 COMPUTER TECHNOLOGY   3
1.3 COMPUTER NETWORK   6
     1.3.1 Local Area Network/Wide Area Network   7
     1.3.2 Internet   9
     1.3.3 Intranet/Extranet   10
     1.3.4 Virtual Private Network   10
1.4 CONSTRUCTION AND COMPUTERS   11
     1.4.1 Features of the Construction Industry   11
     1.4.2 The Construction Process   11
     1.4.3 Computerisation in Construction   13
     1.4.4 IT Usage in Construction Team   15
     1.4.5 IT Construction Best Practice Programme   15
1.5 OVERVIEW OF CONSTRUCTION APPLICATIONS   16
     1.5.1 Business and Information Management   17
     1.5.2 Computer Aided Design and Visualisation   18
     1.5.3 Building Engineering Applications   18
     1.5.4 Computer Aided Cost Estimating   19
     1.5.5 Planning, Scheduling and Site Management   19
     1.5.6 Computer Aided Facilities Management   20
     1.5.7 Integration   21
     1.5.8 Other Types of Applications   21
1.6 ON-LINE RESOURCES   21
SUMMARY   24
DISCUSSION QUESTIONS   24

CHAPTER 2
Business and Information Management

LEARNING OBJECTIVES   25
INTRODUCTION   25
2.1 FUNDAMENTALS OF MANAGEMENT AND INFORMATION   26
     2.1.1 Management Theory   26
     2.1.2 Data, Information and Knowledge   26

2.1.3	Management Information Systems	27
2.1.4	The Latham and Egan reports	28
2.1.5	Professional Office Management	28
2.2	BUSINESS PROCESS REENGINEERING	30
2.2.1	The Five Forces Model	30
2.2.2	Changing Requirements and Business Strategies	30
2.2.3	Business Transformation	32
2.2.4	An Example of Partnering	32
2.3	KNOWLEDGE MANAGEMENT	33
2.3.1	The Conversion of Knowledge	33
2.3.2	Process of Knowledge Management	34
2.3.3	Knowledge Based Systems	35
2.3.4	The Role of IT in KM	36
2.4	INFORMATION MANAGEMENT	38
2.4.1	New Techniques for Managing Information	38
2.4.2	Electronic Document Management	39
2.5	COLLABORATIVE WORKING SYSTEMS	40
2.6	E-BUSINESS	41
2.6.1	Growth Patterns	41
2.6.2	Types of E-business	42
2.6.3	Portals	43
2.7	PROCUREMENT AND MANAGEMENT OF IT SYSTEMS	44
2.7.1	IT System Procurement	44
2.7.2	IT System Management	45
2.8	ON-LINE RESOURCES	46
	SUMMARY	50
	DISCUSSION QUESTIONS	50

CHAPTER 3
CAD and Visualisation

	LEARNING OBJECTIVES	51
	INTRODUCTION	51
3.1	COMPUTER AIDED DESIGN AND DRAFTING	53
3.2	CAD FUNDAMENTALS	54
3.2.1	The Coordinate Systems	54
3.2.2	Drawing Environment Settings	56
3.2.3	Drawing Template	56
3.3	DRAWING WITH CAD	56
3.3.1	Creating Objects	56
3.3.2	Editing	57
3.3.3	Annotations	58
3.3.4	Blocks and External References	58
3.3.5	Grid and Snap	58
3.3.6	Layering	58
3.3.7	File Format	60
3.4	BENEFITS AND LIMITATIONS OF 2D CAD	60
3.5	THREE DIMENSIONAL (3D) MODELLING	62

    3.5.1 Wireframe Model   62
    3.5.2 Surface Model   63
    3.5.3 Solid Model   63
  3.6 VISUALISATION AND ANIMATION   64
    3.6.1 The Visualisation Process   64
    3.6.2 Rendering   65
    3.6.3 Animation   67
  3.7 VIRTUAL REALITY   68
    3.7.1 VR Concept   68
    3.7.2 Different Types of VR   69
    3.7.3 VR Display Systems   69
    3.7.4 VR Input Devices   73
    3.7.5 Virtual Reality Modelling Language   73
    3.7.6 Comparison between VR and Other Visualisation   74
    3.7.7 VR Applications in Construction   75
  3.8 ON-LINE RESOURCES   75
  SUMMARY   78
  DISCUSSION QUESTIONS   78

CHAPTER 4
Building Engineering Applications

  LEARNING OBJECTIVES   79
  INTRODUCTION   79
  4.1 ENGINEERING DESIGN   79
  4.2 BUILDING SIMULATION   80
  4.3 ENERGY ANALYSIS SYSTEMS   81
  4.4 LIGHTING ANALYSIS SYSTEMS   83
  4.5 HVAC DESIGN SYSTEMS   85
  4.6 STRUCTURAL ANALYSIS   86
  4.7 ON-LINE RESOURCES   87
  SUMMARY   90
  DISCUSSION QUESTIONS   91

CHAPTER 5
Computer Aided Cost Estimating

  LEARNING OBJECTIVES   92
  INTRODUCTION   92
  5.1 COST ESTIMATING PRINCIPLES   92
  5.2 UNIT COST ESTIMATING   93
    5.2.1 Quantity Take-off   93
    5.2.2 Labour Costs   94
    5.2.3 Materials, Equipment and Subcontractors   94
    5.2.4 Indirect Costs   95
    5.2.5 Compilation, Analysis and Reporting   95
  5.3 COMPUTER AIDED ESTIMATING   95

5.4 SPREADSHEET APPLICATIONS   96
    5.4.1 Basic Concepts   96
    5.4.2 Constructing Formulas   96
    5.4.3 What-if Analysis   98
    5.4.4 Graphic Outputs   99
    5.4.5 Customised Applications   100
    5.4.6 Evaluation of Spreadsheet   100
5.5 COST ESTIMATING SOFTWARE   102
    5.5.1 Electronic Bill of Quantities   103
    5.5.2 Link BoQ to Cost Libraries   106
    5.5.3 Cost Analysis   107
    5.5.4 Producing Reports   108
5.6 ON-LINE RESOURCES   108
SUMMARY   110
DISCUSSION QUESTIONS   111

CHAPTER 6
Planning, Scheduling and Site Management

LEARNING OBJECTIVES   112
INTRODUCTION   112
6.1 CONCEPTS OF PROJECT PLANNING AND SCHEDULING   112
    6.1.1 Top Down Planning   112
    6.1.2 Work Package Structure (WPS)   113
    6.1.3 Programming and Scheduling   113
    6.1.4 Resource Allocation   115
6.2 BUILDING A PLAN USING PROJECT PLANNING PROGRAMS   115
    6.2.1 Define a Project   116
    6.2.2 Enter a Task List   117
    6.2.3 Schedule Tasks   118
    6.2.4 Assign Resources   119
    6.2.5 Optimise a Plan   120
6.3 TRACK AND MANAGE A PROJECT   121
    6.3.1 Manage the Schedule   121
    6.3.2 Manage Work   121
    6.3.3 Track Costs   122
    6.3.4 Balance Workload   123
6.4 SITE MANAGEMENT   124
    6.4.1 Operation Simulation   124
    6.4.2 Site Monitoring   125
    6.4.3 Material Procurement   125
6.5 MOBILE COMPUTING AND ITS ON-SITE APPLICATION   126
    6.5.1 Wireless Communication   126
    6.5.2 Applications On-Site   127
6.6 ON-LINE RESOURCES   128
SUMMARY   130
DISCUSSION QUESTIONS   131

CHAPTER 7
Computer Aided Facilities Management

LEARNING OBJECTIVES 132
INTRODUCTION 132
7.1 THE ROLE OF FACILITIES MANAGEMENT 133
   7.1.1 Core and Non-core Business Activities 133
   7.1.2 Growth of FM 133
7.2 COMPUTER AIDED FM 134
   7.2.1 Helpdesk 135
   7.2.2 Asset Management 136
   7.2.3 Space Management 136
   7.2.4 Building Operation Management 137
   7.2.5 Other Functions 138
7.3 ON-LINE HELPDESK SYSTEM 138
   7.3.1 Deficiencies of Stand Alone Helpdesk Systems 138
   7.3.2 On-line Systems 140
7.4 FUTURE TRENDS 143
   7.4.1 Flexible Working 143
   7.4.2 The Future of Facilities Management 144
7.5 INTELLIGENT BUILDINGS 144
7.6 ON-LINE RESOURCES 145
SUMMARY 147
DISCUSSION QUESTIONS 147

CHAPTER 8
Integration

LEARNING OBJECTIVES 148
INTRODUCTION 148
8.1 THE NEED FOR INTEGRATION 149
8.2 METHODS OF INTEGRATION 150
8.3 PRODUCT AND PROCESS MODELS 152
   8.3.1 Modelling 152
   8.3.2 Product Models 152
   8.3.3 Process Models 153
8.4 STANDARDS 154
   8.4.1 Data Model Standards 154
   8.4.2 Data Modelling Standards 155
   8.4.3 Data Mapping Standards 155
   8.4.4 Data Exchange Standards 156
8.5 INTEGRATED PROJECT DATABASES 157
8.6 AN EXAMPLE INTEGRATED SYSTEM – GALLICON 158
   8.6.1 The GALLICON System 159
   8.6.2 Design Collaboration Using GALLICON 161
   8.6.3 Benefits of the Integrated System 162
8.7 ON-LINE RESOURCES 164

SUMMARY   165
DISCUSSION QUESTIONS   165

CHAPTER 9
Future Developments

LEARNING OBJECTIVES   166
INTRODUCTION   166
9.1 TECHNOLOGY FORESIGHT   166
    9.1.1 High Speed Network and Mobile Network   167
    9.1.2 Pervasive or Ubiquitous Computing   169
    9.1.3 Intelligent Speech Recognition   169
    9.1.4 Telepresence   170
    9.1.5 Computing Grid   170
    9.1.6 Distributed Virtual Environments   172
    9.1.7 Adaptive and Learning Systems   172
9.2 ACCELERATING CHANGE IN CONSTRUCTION   173
9.3 VISION OF CONSTRUCTION IT   174
    9.3.1 Questions for a Company Developing a Vision 174
    9.3.2 A Vision for the Construction Industry   175
    9.3.3 Barriers to Success   175
    9.3.4 Stages in the Development of a Vision   176
9.4 EMERGING TRENDS   176
9.5 ON-LINE RESOURCES   177
CONCLUSION   179
DISCUSSION QUESTIONS   179

REFERENCES   180
INDEX   184

FIGURES

Figure 1.1   Common LAN topologies   8
Figure 1.2   The Internet architecture   9
Figure 1.3   Computer performance improvement and increase of computer use   14
Figure 1.4   Construction process and IT applications   17
Figure 2.1   Porter's five forces model (after Porter, 1980)   30
Figure 2.2   Business process reengineering   31
Figure 2.3   Heathrow Terminal 5 model
(reproduced with the kind permission of BAA)   32
Figure 2.4   Knowledge conversion process (after Nonaka, 2000)   33
Figure 2.5   Knowledge management using Case Based Reasoning
(after Bush, 1999)   35
Figure 2.6   Configuration of an electronic document management system   39
Figure 2.7   The rise and fall of e-business companies
(after Gartner Group, 2001)   41
Figure 3.1   2D drawing (image courtesy of Autodesk)   52
Figure 3.2   Multiple views of building produced by CAD   54
Figure 3.3   Coordinate system   55
Figure 3.4   Basic 2D objects   57
Figure 3.5   The use of layering in 2D drawing   59
Figure 3.6   Wireframe model   62
Figure 3.7   Surface model   63
Figure 3.8   Solid model   64
Figure 3.9   A computer generated picture
(image courtesy of IDPartnership-Northern)   65
Figure 3.10 A VR model (image courtesy of Salford VE Centre)   67
Figure 3.11 Sample head-mounted display devices
(images courtesy of Cybermind Interactive Nederland)   70
Figure 3.12 BOOM (image courtesy of Fakespace Systems Inc.)   70
Figure 3.13 Reality Centre (image courtesy of Fakespace Systems Inc.)   71
Figure 3.14 Workbench (image courtesy of Fakespace Systems Inc.)   72
Figure 3.15 Illustration of a CAVE system (image courtesy of
University of Michigan Virtual Reality Laboratory)   72
Figure 3.16 Data glove and its interaction with a car VR model (images
courtesy of University of Michigan Virtual Reality Laboratory)   73
Figure 3.17 A VMRL model   74
Figure 4.1   Condensation prediction within external wall
(image courtesy of Owens Corning ACS)   81
Figure 4.2   Sources of heat gains over 24 hours
(reproduced with the kind permission of Hevacomp Ltd)   82
Figure 4.3   Lighting analysis (image courtesy of Cymap)   84
Figure 4.4   Mechanical design showing visualisation and drafting of ductwork
(reproduced with the kind permission of Hevacomp Ltd)   85
Figure 4.5   Sydney Opera House – shells designed using structural analysis
software by Ove Arup & Partners, engineers. Jørn Utzon, architect   86
Figure 5.1   Spreadsheet user interface   97
Figure 5.2   Excel pie chart   99

Figure 5.3   User interface of a bespoke spreadsheet program   101
Figure 5.4   A bespoke spreadsheet program   101
Figure 5.5   Flowchart of the use of a cost estimating program   103
Figure 5.6   A graphic take-off program (image courtesy of Databuild) 105
Figure 5.7   Dialog box for entering new bill items
             (image courtesy of Masterbill)   105
Figure 5.8   Cost estimate print out
             (image courtesy of Worldwide Software (UK) Ltd)   108
Figure 6.1   A simple PERT chart   114
Figure 6.2   A CPM chart   114
Figure 6.3   A Gantt chart   115
Figure 6.4   Start a new project schedule file   116
Figure 6.5   Task list window   117
Figure 6.6   Schedule and task dependencies   118
Figure 7.1   Maintenance job process using helpdesk system   135
Figure 7.2   Space management system
             (image courtesy of www.cafmexplorer.com)   137
Figure 7.3   An on-line helpdesk system   141
Figure 7.4   Maintenance job process using knowledge based
             on-line helpdesk system   142
Figure 8.1   Total integration around a single project database   151
Figure 8.2   Communications between individual applications
             that can be linked gradually   151
Figure 8.3   The evolution of data exchange standard (after Warthen, 1989)   154
Figure 8.4   An example STEP file   157
Figure 8.5   System architecture of GALLICON   159
Figure 9.1   The VIRTUE telepresence station
             (images courtesy of the BT's VIRTUE project)   171
Figure 9.2   Rethinking Construction improvement targets   174

TABLES

Table 1.1	Timeline of computing since 1951	4
Table 2.1	Three main schools of management theory (after Laudon, 2000)	26
Table 2.2	Traditional and contemporary management (after Laudon, 2000)	27
Table 2.3	Decision making and information requirements	29
Table 2.4	Four generic strategies (after Porter, 1980)	31
Table 2.5	The impact of information technology on knowledge management (after Davenport, 1993)	36
Table 2.6	Internet business models (after Laudon, 2000)	42
Table 3.1	An example to illustrate a layer name using all fields of the British standard	59
Table 5.1	Commonly used spreadsheet functions	98
Table 9.1	Network capacity comparisons	168

PREFACE

In recent years, the use of computers in the UK construction industry has been growing steadily. Information Technology (IT) is widely regarded as a key enabler for further performance improvement in the industry. IT knowledge and computer skills have become essential requirements for new graduates. IT related modules are an integral part of the curriculum of building and construction undergraduate courses. There is also an increasing number of postgraduate courses related to the subject of IT in construction. The requirement for construction IT education is not just to teach students how to operate CAD and some other application packages. It is important to equip them with a broad knowledge of what, where and how IT applications are used, or can be used, during different stages of the whole life cycle of buildings. The students need to know the functionality as well as the business benefits of the main types of IT applications in construction.

This book is written for this purpose. It seeks to provide an introduction to several main types of IT applications currently used in construction. It is suitable for use as a textbook for all students of AEC (Architecture, Engineering and Construction) related courses. Professionals in the construction industry will also find it useful as a reference book.

Recognising the fast changing nature of the subject matter, we have included a large number of references to external information. A collection of web links is provided in each chapter. These are not meant to be exhaustive. Nevertheless, they provide a good source of extra information. Several case studies are also included in the book. They are successful examples of where IT applications are used to the benefits of the business process.

The book starts with an overview of existing computer technology and the current use of IT in construction in Chapter 1. Chapters 2-7 address six types of the most widely used construction IT applications in a logical sequence using the construction process model as a framework. Chapter 8 discusses the integration between different applications. The final chapter looks at the future development of computer technology and its potential applications in construction. For each chapter a set of *Learning Objectives* is outlined at the beginning. They give an indication of the context of that chapter and the expected outcomes after studying it. At the end of each chapter, a list of *Discussion Questions* is given. Readers can used these as a checkpoint and for personal feedback after studying the materials.

The book is not intended as a tutorial guide for learning to use individual software. The reference to specific software packages is entirely for the purpose of demonstrating the main functionality of each type of IT application. It does not imply any endorsement of the packages by the authors.

ACKNOWLEDGEMENTS

The book originated from the authors' teaching materials for the MSc in Construction IT course at the University of Salford. The course was developed with funding from EPSRC and support from a consortium of construction companies. Without their contribution, this book would not have been started.

We wish to thank Dr John Connaughton of IT Construction Best Practice programme and Paul Wilkinson of BIW Technologies for allowing case study materials to be used. We thank the Construction Industry Computing Association for permission to make use of their Software Directory.

We have used some materials, mainly pictures, reproduced from other sources. We have acknowledged their original sources and give our thanks to the copyright holders for their permission for us to use these images. The following figures used 'screen shots reprinted by permission from Microsoft Corporation': Figures 5.1, 5.2, 5.4, 6.3, 6.4, 6.5 and 6.6.

ABBREVIATIONS

ADSL	Asymmetric Digital Subscriber Line
ARPANET	Advanced Research Project Agency Network
BMS	Building Management System
BOOM	Binocular Omni-Orientation Monitor
BoQ	Bills of Quantities
BSCW	Basic Support for Collaborative Work
CAD	Computer Aided Design
CAFM	Computer Aided Facilities Management
CAVE	Cave Automatic Virtual Environment
CBR	Case Based Reasoning
CD-RW	Compact Disc Rewriteable
CICA	Construction Industry Computing Association
CMC	Computer Mediated Communication
CPM	Critical Path Method
CPU	Central Processing Unit
CRM	Customer Relations Management
CSCW	Computer Supported Co-operative Work
DVD	Digital Versatile Disc
EDM	Electronic Document Management
EGA	Enhanced Graphics Adapter
ERP	Enterprise Resource Planning
FM	Facilities Management
FTP	File Transfer Protocol
HMD	Head Mounted Display
HTML	Hype Text Markup Language
HVAC	Heating, Ventilating and Air Conditioning
IAI	International Alliance for Interoperability
IFC	Industry Foundation Class
I/O	Input/Output
ISDN	Integrated Services Digital Network
ISP	Internet Service Provider
KBS	Knowledge Based System
LAN	Local Area Network
LCD	Liquid Crystal Display
MIS	Management Information System
OCR	Optical Character Recognition
PDA	Personal Digital Assistant
PC	Personal Computer
RAM	Random Access Memory
SMM	Standard Methods of Measurement
STEP	Standards for the Exchange of Product data
UCS	User Coordinate System
UNIVAC	Universal Automatic Computer
VGA	Video Graphics Adapter
VPN	Virtual Private Network
VR	Virtual Reality

VRML	Virtual Reality Modelling Language
WAN	Wide Area Network
WCS	World Coordinate System
WPS	Work Package Structure
WWW	World Wide Web
XML	Extensible Markup Language

CHAPTER 1

Overview of IT Applications in Construction

LEARNING OBJECTIVES

1. Appreciate the impact of IT on modern society.
2. Gain knowledge of computers and computer networks.
3. Understand the development of computers in construction.
4. Recognise the major categories of current IT applications and their use in the construction process.

INTRODUCTION

Computing and communication technology, also commonly known as Information Technology (IT), have been radically transforming the way we live, learn, work and play (Capron, 2000). Although it is widely acknowledged that the construction industry is lagging behind other industries in adopting IT (Department of Environment, 1995), the penetration of computers in construction has been gathering pace in recent years due to rapid improvements in computer hardware and software (Stevens, 1991; CICA, 1993; CICA, 1996; Building Centre Trust, 1999). Today, a large number of software packages are available to all disciplines of the construction team at all stages of the construction process. They provide support for a broad range of activities such as computer aided design and drafting, building visualisation, design appraisal, project management, information storage and retrieval, cost estimating, structural analysis, on-site management, facilities management and others.

In this chapter, we begin with an overview of the impact of computers on modern society and a review of the progress of computerisation in the construction industry. Then we briefly discuss the major types of IT applications to provide a context for more detailed discussions in the later chapters.

1.1 THE INFORMATION AGE

The impact of IT on modern society is profound. It is often described as the 'Information Revolution'. The driving force behind this revolution is the convergence of computing and telecommunication technology. The changes IT has brought about to modern society are far beyond just automation of some routine tasks. In many cases, IT has fundamentally changed the way business is conducted. The pace of change is also unprecedented. Previous social revolutions such as the industrial revolution have taken a long time, hundreds of years, to spread out from their original sources, but the 'Information Society' emerged all around the world and within a single generation. Today, Internet access has spread to even the least developed regions of the globe.

IT has enabled the globalisation of the economy and competition, and has subsequently brought about large-scale changes in the industrial makeup of all the advanced industrial nations. We have witnessed the rapid growth of some industries such as computers, communications, software and financial services by enabling new products, services, and efficiencies, while other more traditional industries have stalled or even contracted by comparison.

As it is transforming the economic landscape, IT is also bringing a major shift in the job market. Many analysts have noted that information technology is resulting in a more polarised occupational structure, consisting of highly skilled, well paid jobs at one end and lower skilled, low wages at the other, with few jobs in between. For the modern work force, IT literacy is becoming an essential requirement. With the arrival of digital television, on-line shopping and banking, IT becomes a basic skill every member of the community needs to have.

Today computers are everywhere, in offices, stores, banks, homes and even coffee shops. The following are only a few examples of where computers are used.

Communication: The ability to communicate underpins most human activities. Traditional communication media include telephone, fax and mail. In the new information era, people can exchange correspondence almost instantly using e-mail, access unlimited information on the Internet and meet other people over physical distances using video conferencing. The new communication technology enables people located in different places to work together as if they were in the same office. Big multinational companies are already exploiting this technology to achieve better use of the resources of their separate offices. For example, design may be done in one country while construction may be carried out in another. In this way, projects can be shared between offices so that the best expertise can be applied to the job. Furthermore, companies with international offices can now work around the clock. When workers in one office finish work at the end of the day, files can be passed on to an office in a different time zone for work to continue.

On-line services: The rapid development of the Internet and the World Wide Web (WWW) has enabled many services that traditionally require face-to-face contacts to be delivered on-line. For example, going to the local bank branch used to be the normal way for people to carry out financial transactions. Now, on-line banking allows customers to manage their own bank accounts 24 hours a day in the comfort of their own homes. People can pay bills, transfer money, and set up standing orders over a secure Internet link without the need to trouble a bank clerk. Similarly, many supermarkets also provide on-line shopping services. Customers can browse and order goods on-line. The shops will deliver the goods to the customers' homes.

Internet distance learning: Providing more people with opportunities of university education is important to a country's competitiveness in a global market. Widening access to higher education includes delivering teaching and learning to people who cannot attend lectures at the campus. The traditional correspondence based distance learning does not allow close interaction between students and tutors. Today, computers and the Internet allow students to study using an on-line virtual learning environment. In a similar way to a campus environment, students can interact with tutors and fellow students. They can access course materials and library services. They can even carry out on-line assessments. This new development is particularly beneficial to those people who are in work. Now they can study in their own time from home or workplace.

E-business: In addition to the above business-to-customer on-line services, there are more and more business-to-business services. The Internet provides a virtual market place for buyers, suppliers, distributors and sellers to exchange information, negotiate with one another and conduct trade. This development brings new market opportunities for many companies because trade is no longer restricted by physical distance. On the other hand, there will be more competition as many companies can potentially participate in a given business transaction.

Teleworking: As IT brings rapid changes in the workplace, it also brings dramatic changes in the way people work. With Internet access from home and public places, many workers no longer need to come to the office daily. They can access a company's computer servers at any time and any place to download their assignments and keep in contact with fellow workers. This gives more flexibility for the individuals, especially for those who need to combine work with family responsibilities. Companies benefit from the need for less office space and, very often, more productive workers. The whole society also benefits from less traffic on the roads.

1.2 COMPUTER TECHNOLOGY

A computer is a complex machine that can process information following predefined instructions. The creation of computers can be attributed to people's need to do calculations in daily life and their desire to seek the help of tools. The Chinese invented the abacus 5000 years ago, and it remained in wide use in China until the 1970s. In Europe, the early pioneers of mechanical calculation devices include Blaise Pascal (1623-1662), Gottfried Wilhem von Leibniz (1646-1716), and Charles Babbage (1791-1871). In the middle of the 20th century, computers entered the digital age.

The first generation (1945-1956) digital computers were purpose built machines. Each of them had a specific program of instructions. Once they were built, it was difficult to program them for a different purpose. They used bulky vacuum tubes, which made the computers very large. Most of these computers would occupy a whole room. The second generation (1956-1963) used transistors. Transistors are much smaller than vacuum tubes so that more of them can be packed into one computer, and computers became smaller and faster. During this period, high level programming languages, such as COBOL (Common Business-Oriented Language) and FORTRAN (Formula Translator) came into common use. As a result, computers became programmable by their users. During the third generation (1964-1971) of computers, Integrated Circuits and semiconductors replaced transistors. These new components were more reliable and lasted longer. Computer operating systems emerged and many third party software packages could be installed and run on these computers. The fourth generation (1971-present) still uses integrated circuits. However, due to the improvement in manufacturing technology and the use of new materials, the size of the circuits becomes miniaturised and millions of components can be fitted on to a small chip. The cost of these chips also dropped rapidly. Computers were no longer exclusive to large businesses. They became affordable for small businesses and individuals. In addition to the shared mainframe computers, Personal Computers (PCs) have become more widely used in offices and homes.

The following table is a brief timeline of computing development since the first commercial computer in 1951.

Table 1.1 Timeline of computing since 1951

Year	Event
1951	UNIVAC (Universal Automatic Computer), the first commercial digital computer was delivered to the US Bureau of Census.
1954	FORTRAN (FORmula TRANslation) programming language was developed. It, together with the later COBOL (1959) and BASIC (1965), allowed programmers to write software using high level, more English- like languages.
1969	ARPANET (Advanced Research Projects Agency Network) was started by the US Department of Defence. The network established links between several computers located in different places. It was the origin of today's Internet.
1974	MITS Altair 8800, the first Personal Computer (PC) was produced. It was a gee-whiz machine with switches and dials, but no keyboard or monitor.
1975	Microsoft was formed to provide operating system for IBM PCs. Since then, it has developed MS-DOS, MS Windows, Windows NT, Windows 95, Windows 2000. Today, over 90% of the PCs have Microsoft operating systems installed.
1981	MS-DOS 1.0 released.
1981	IBM PC was released. The standard model was equipped with 64k of RAM, a monochrome monitor and two optional, single sided floppy drives. IBM defined the architecture standard for PCs.
1983	IBM XT released. It had an 8086 processor, 10MB of hard disk, 128k of RAM, one floppy drive, and monochrome monitor.
1985	The CD-ROM was invented. It can store hundreds of megabytes of data.
1985	MS Windows was launched. It provided a graphic user interface, with windows, dialog boxes, menus, icons, etc. However, Windows versions 1 and 2 were not widely used. The release of Windows 3.0 in 1990 made Microsoft the standard operating system for PCs.
1986	IBM introduced its first laptop computer. Since then, many manufacturers have been producing portable computers that people can use while on the move.
1986	Intel released the 80386 processor. A standard 386 PC would have an Intel 20MHz 80386 processor, 2MB of RAM, 40MB of hard disk

1989	Intel release the 80486 processor. A standard 486 PC would have an Intel 33MHz 80486 processor, 16MB of RAM, 250MB of hard disk.
1989	The World Wide Web (WWW) was invented at CERN (European Particle Physical Laboratory). The WWW uses the Internet as a vehicle to link up distributed digital resources on different computers through hyperlinks. It presents the users with useful information regardless of where it is stored. The initial WWW had a text-based interface. Later, the NCSA Mosaic browser provided a graphic interface for it and made it accessible to novice users. As a result, its popularity exploded and WWW has become synonymous with the Internet.
1993	Intel released the Pentium processor. A standard Pentium PC would have an Intel 100MHz Pentium processor, 32MB of RAM, 1.2GB of hard disk.
1995	Microsoft Windows 95 released. It is a complete operating system replacing the MS-DOS and MS Windows 3.X combination. It provides plug and play support so that many accessories, such as printers, network and modem cards, scanners and digital cameras, can be installed easily. It has been very popular with users.
1995	The Palm Pilot, a pen-based Personal Digital Assistant (PDA) designed for portability, was launched by Palm Computing. Initially, it was aimed at providing a digital diary for taking notes, managing contact details and appointments. The latest PDA can synchronise with PCs, access the Internet, and provide ample processing power for normal office applications.
1996	56k Modem produced. A modem is a device that converts digital signals to analogue ones and vice versa. It enables computers to communicate with each other through telephone wires. Although the modem was first invented in the 1950s, it was the 56k modem that connected many homes to the Internet.
2000	Microsoft Windows 2000 released.
2002	A high specification PC would have a 2.4GHz Pentium 4 processor, 256MB of RAM, 80GB of hard disk, a 17 inch, 16.7 million colour, high resolution LCD monitor, DVD and, CD-RW drives, multimedia capability, network enabled. Such a PC still costs a lot less than a standard PC did two years before.

What make computers so useful are their fast data processing speed, massive storage capacity, reliability, and choice of input/output devices.

Speed: A computer processes information extremely quickly, from several thousand to several billions of instructions per second. Without computers, some tasks, such as space exploration, missile precision targeting, could not possibly be carried out. Real time computer systems are able to respond to the instructing input with a minimum of delay. The fast processing speed can help to increase the productivity of any task and improve the man/machine interaction.

Massive storage: During the past few decades, there has been a massive increase in information. At present, most of the tasks in scientific research, office work and building design involve handling large amounts of information in various forms, i.e. text, numbers or graphics. The computer provides an excellent means of storing and retrieving such information.

Reliability: Computers are extremely reliable at doing what they are told. It is true that sometimes computers have failures or 'crashes' during operation. However, these failures are usually caused, not by the computer itself, but by human error during either the programming stage or in use.

Multiple interface devices: There is a wide variety of I/O devices available such as keyboard, scanner, microphone, monitor, printer, plotter, etc., which can be used to read and present text, graphics and sound. The combination of the multiple I/O devices offers a rich capacity for developing application software.

Today, computer performance is measured in billions of instructions per second. Palm-top computers far exceed the performance of the 1960s mainframes at a fraction of the price. Wireless technology and the integration of data, voice, video and graphics capabilities over very fast, yet cost-effective, networks now allow 'any time, any place, and any form' communication and information sharing. These dramatic improvements in IT price/performance will lead to equally dramatic changes in organisational strategy, structure, processes, distribution channels and work.

1.3 COMPUTER NETWORK

A computer network is two or more computers being linked together so that data can be exchanged between them. Computer networks enable people to send messages to the other side of the world instantly at the click of a button. They also allow IT resources, such as computer servers, printers, and scanners, to be shared by all the employees in an organisation. The basic components of a computer network include:

1. A sending device
2. A communication link
3. A receiving device.

The sending device is usually a computer. The receiving device can be other computers or network printers. The communication link is the physical medium through which communication signals can be transmitted. The common types of link include copper wire, coaxial or twisted pair cable, and optical fibre. There are also wireless connections, such as infra red, radio wave and satellite links. These different types of link have different capacity and transmission speed. For example, twisted pair cable is faster than copper wire, and optical fibre is faster still. However, faster links are usually more expensive to install. Therefore, each organisation should choose its network links on the basis of a cost benefit analysis.

In practice, most computer networks have more complex configurations. The main types of networks and their configurations are discussed in the following.

1.3.1 Local Area Network/Wide Area Network

A Local Area Network (LAN) usually connects a group of computers in close proximity to each other such as in a building, a department or a small organisation. The main aim of a LAN is usually to share resources, such as printer and file server. The layout of a network is called topology. Figure 1.1 shows the commonly used LAN topologies.

- In a *Ring topology*, the main cable forms a closed loop. All computers connect to the loop as a node. The advantages of this configuration include: (1) the need for less cable hence less expensive; and (2) no need for central wiring closet. However, its disadvantages are: (1) a single node failure could cause the whole network to fail; (2) it is difficult to diagnose and locate a fault when it occurs; (3) it is difficult to modify the layout of the network once it is installed.
- In a *Start topology*, all computers are connected through a central hub. The advantages of this configuration are: (1) cable layouts are easy to modify; (2) a workstation can be added easily; and (3) it is easy to identify network break down. Its disadvantages include: (1) it requires large amounts of cable hence it is more expensive to set up; and (2) it has a crowded central hub especially when the number of computers is large.
- In a *Linear Bus topology*, all computers are connected to a single cable that runs from one end to another. This configuration requires the least amount of cable. Therefore, it is usually the least expensive option. It is also easy to extend. However, the main trunk is often a communication bottleneck. It is also most fragile. Any fault along the main cable will cause a break down.
- In addition to the above three topologies, there are many other *hybrid configurations* that combine the three basic topologies.

A Wide Area Network is a network that is spread across a wide geographic area. Typically a WAN consists of two or more LANs. It is usually built to provide communication solutions for organisations or people who need to exchange digital information between distant places. For example, a large company might have offices in different locations within a country or even in different countries. A WAN will enable all its departments and employees to communicate and share information easily and securely even though they may be located long distance apart.

The communication infrastructure for WANs is usually provided by public telecommunication service providers, such as telephone or cable companies. Other organisations who need to set up a WAN can lease lines from these service providers. The lease cost varies depending on the required communication capacity and complexity of network configuration.

In the UK, Super JANET (Joint Academic Network) is a WAN linking all universities and colleagues in the country. The Internet is the largest WAN in the world. Further discussion is available in the next section.

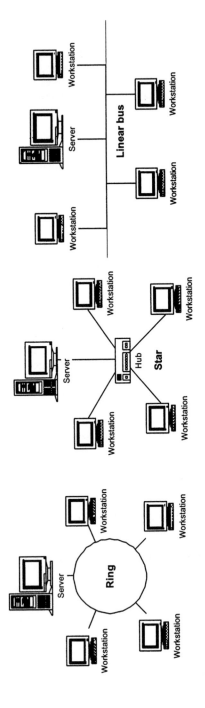

Figure 1.1 Common LAN topologies

1.3.2 Internet

The Internet originated from the Advanced Research Projects Agency Network (ARPANET) of the Department of Defense in the United States. ARPANET was a project started in the late 1960s to provide a test-bed for emerging network technologies. Initially, it only connected four universities in the US. It has expanded rapidly since the 1970s. Today hundreds of millions of computers worldwide are connected to the Internet. Figure 1.2 illustrates the basic architecture and the main components of the Internet.

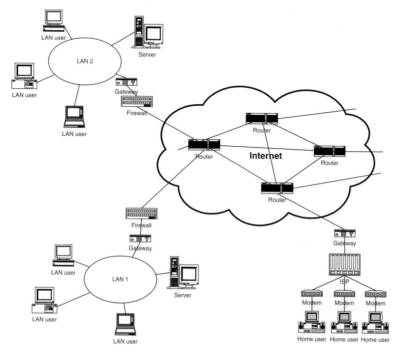

Figure 1.2 The Internet architecture

On the Internet, each computer has a unique identity called an Internet Protocol (IP) address. TCP/IP is a set of protocols developed to allow computers to share data across the Internet. When a file is transferred from one computer to another, the data of the file is broken into small packets. Each packet is marked with the identity of the sender and receiver and its position in the sequence of packets for the whole message. The packets travel on the Internet separately to their destination. Different packets may use different routes. Routers are devices that control the traffic of packets on the Internet. When all packets arrive at their destination, they are reassembled in the correct order to form the original file.

Computers within a LAN usually do not connect to the outside Internet directly. Their connection to the Internet is through a Gateway that has links to the external gateways of other networks.

A firewall is a device or software application that serves as a flexible barrier that sits between the computers on a LAN and the outside world (i.e. the Internet). Firewalls apply a set of rules to decide who is allowed to connect to which machines and what services they are authorised to use. A firewall, when set up properly, provides an excellent means for protecting a network, and the machines connected to it, from intrusion. A firewall's primary purpose is to guard an internal network against malicious access from the outside; however, firewalls may also be configured to limit access to the outside from internal users.

Home users, who do not have access to a LAN, can use the service offered by Internet Service Providers (ISPs). An ISP is a company that provides Internet connectivity to home and business customers. Home users need a modem device to dial up a connection to the ISP's server using the public telephone system. The transmission speed of a modem is not very fast. Users who need faster connections can use Integrated Services Digital Network (ISDN) or Asymmetric Digital Subscriber Line (ADSL). However, ISDN and ADSL are more expensive than the normal modem connection. The latest development of networking technology is discussed in Chapter 9.

The popularity of the Internet is largely due to its useful applications. The World Wide Web (WWW), developed by Tim Berners-Lee in 1993, is the most widely used application so much so that it has almost become synonymous with the Internet itself. The WWW is a hyper linked database of billions of documents hosted on millions of servers on the Internet. Ordinary users can view and access these documents using user-friendly browsers without worrying where they are actually stored. The WWW has been essential for the rapid growth of commercial use of the Internet in recent years.

In addition to the WWW, there are several other widely used applications, including e-mail, File Transfer Protocol (FTP), USENET news groups, Chat & Instant Messaging, etc.

1.3.3 Intranet/Extranet

An Intranet, also known as a private network, is an application of a LAN. It uses standard Internet technologies, such as TCP/IP, WWW servers and browsers, within an organisation to allow its employees to find, use, and share documents and communicate with each other. An Intranet usually has controlled access to the Internet behind a firewall. No access from outside is allowed.

An Extranet is a computer network that allows controlled access from the outside for specific business purposes. It is an extension to an Intranet by allowing trusted customers and/or business partners to connect to the otherwise private Intranet usually via the Web. The access controls are set at the firewall.

1.3.4 Virtual Private Network

The Internet is a public network. It is convenient to use but not secure. Traditionally organisations that wanted to build a wide-area network needed to procure expensive, dedicated lines to connect their offices together. Very few companies could afford to purchase these lines outright. Most organisations 'leased' their lines and paid a monthly charge for the privilege of using cables as a private network that no one else could tap into.

A Virtual Private Network (VPN) uses the Internet as the transport backbone to establish secure links between business partners, extend communications to regional and isolated offices. The key to VPN is the encryption technology. Before a piece of data is transferred to the Internet, it is encrypted. The intended receiver, and only the intended receiver, has appropriate keys to decrypt the message and reveal the original data. Even if the message is intercepted by others on the Internet, they will not be able to work out the real content of the message. A VPN creates a secure virtual communication 'tunnel' on the public Internet. It functions like a private network at a fraction of the cost.

1.4 CONSTRUCTION AND COMPUTERS

1.4.1 Features of the Construction Industry

The construction industry is a major economic sector in the UK, accounting for about 8-10% of the Gross Domestic Product and employing over a million workers. The construction process is often a lengthy and complex one, involving clients, architects, consultants, contractors, and suppliers.

The construction industry is highly fragmented. There are some 163,000 construction companies listed on the then Department of the Environment, Transport and the Regions' (DETR) statistical register (Egan, 1998). Although there are some large multi-national building contractors and clients, the majority of firms in the industry are small or medium sized companies. A typical construction project usually involves an ad hoc team of different firms, each of which only deals with certain aspects of the project. Very often, each firm is only interested in improving its own productivity. This is one of the reasons why many of the current IT applications are directed at a single activity, such as drawing production or cost estimation.

The construction process usually results in a unique product, a specific building for a specific context of site conditions and client requirements. On one hand, this requires a detailed evaluation and appraisal of building performance during the design stages where IT applications can be of great help. On the other hand, design consultants are reluctant to invest in learning the skills required for IT systems and collecting data for operating the systems, when the benefits are perceived as limited.

1.4.2 The Construction Process

The construction process here refers to the whole life cycle of a building, from conception, to design, construction and maintenance. The RIBA Plan of Work (RIBA, 2000) is a well-recognised process protocol that describes the construction process in the following eleven stages:

A. *Appraisal*: Identification of Client's requirements and possible constraints on development. Preparation of studies to enable the Client to decide whether to proceed and to select a probable procurement method.

B. *Strategic Briefing*: Preparation of a Strategic Brief by, or on behalf of, the Client confirming key requirements and constraints. Identification of procedures, organisational structure and range of Consultants and others to be engaged for the Project.

C. *Outline Proposals*: Commence development of a Strategic Brief into a full Project Brief. Preparation of Outline Proposals and estimate of cost. Review of the procurement route.

D. *Detailed Proposals*: Complete development of the Project Brief. Preparation of Detailed Proposals. Application for full Development Control approval.

E. *Final Proposals*: Preparation of final proposals for the Project sufficient for co-ordination of all components and elements of the Project.

F. *Production Information*: Now in two parts, F1: Preparation of production information in sufficient detail to enable a tender, or tenders, to be obtained. Application for statutory approvals. F2: Preparation of further production information required under the building contract.

G. *Tender Documentation*: Preparation and collation of tender documentation in sufficient detail to enable a tender or tenders to be obtained for the construction of the Project.

Additional Information
Rethinking Construction and Accelerating Change

In 1998, Sir John Egan led a Construction Task Force and conducted a study of the UK construction industry. The findings were published in the Rethinking Construction report. He demanded sustained improvements in quality and efficiency. Partnering was recommended to replace competitive tendering to procure construction projects. Integrated processes and teams were considered one of the key drivers of change.

Following the Egan report, the Movement for Innovation (M4I) was established to help the construction industry to respond to the Rethinking Construction agenda. A series of demonstration projects were selected as show cases to promote change and innovation in the construction industry.

In 2002, a follow-up Accelerating Change report was published, which set out specific targets for reduction in cost, construction time, defects and accidents, and for increases in predictability, productivity, and turnover and profits. The report recognised IT and the Internet as important enablers.

Additional information can be found at the following web sites.

www.rethinkingconstruction.org
www.dti.gov.uk/construction
www.m4i.org.uk

H. *Tender Action*: Identification and evaluation of potential Contractors and/or Specialists for the construction of the Project. Obtaining and appraising tenders and submission of recommendations to the Client.

I. *Mobilisation*: Letting the building contract, appointing the contractor. Issuing production information to the contractor. Arranging site handover to the contractor.

J. *Construction to Practical Completion*: Administration of the building contract up to, and including, practical completion. Provision to the Contractor of further Information as and when reasonably required.

K. *After Practical Completion*: Administration of the building contract after practical completion. Making final inspections and settling the final account.

These eleven stages can be divided into five phases:

1. *Requirement Analysis* phase includes Appraisal and Strategic Briefing.
2. *Design* phase covers the Outline Proposals, Detailed Proposals, Final Proposals and Production of Information stages.
3. *Tendering* phase covers the Tender Documentation and Tender Action stages.
4. *Construction* phase covers the Mobilisation and Construction to Practical Completion stages.
5. *Maintenance* phase covers the After Practical Completion stage.

These simplified phases will be used to map out the current IT applications in the construction process later in this chapter.

1.4.3 Computerisation in Construction

For construction professionals, the initial surge of enthusiasm for computer applications started in the early 1960s. There was an optimistic view of the computer's potential as a supporting tool for design and construction and the time needed to develop this potential (Stevens, 1991). By the early 1980s, the initial excitement had been replaced by a greater realism about what the computer could offer. The change of opinion was caused by a combination of the high capital cost of computing hardware and the limitations of the existing software systems. In 1985, an AIA survey in the United States revealed that many architectural firms did not make use of the computer because they believed 'it would cost too much' (Architectural Record, 1985). On the computer performance aspect, there were the following constraints:

1. The small storage capacity that limited the amount of information that could be stored.
2. The processing speed was relatively slow. A simple simulation could take hours. It was impractical to conduct what-if analysis.

3. The I/O facilities were inadequate. The poor display devices were not suitable for handling high resolution graphics needed for design support.

Since the mid 1980s, the penetration of computers in the construction industry has been accelerating thanks to the rapid development in computer hardware and software as discussed in the sections 1.2 and 1.3. Figure 1.3 shows the increase of computer use in construction in parallel with computing performance improvement.

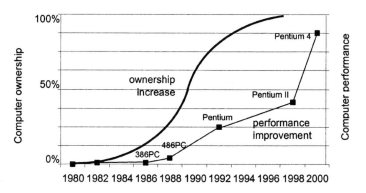

Figure 1.3 Computer performance improvement and increase of computer use

The steady increase of computer ownership and IT applications among construction practices has been confirmed by a series of surveys by RIBA, CICA and other professional organisations (RIBA Journal, January 1995; CICA, 1993). The growing importance of computers is also indicated by the steady increase of R&D activities in this area. Professional magazines, such as Building and the Architects' Journal, have papers on the development of new computer software in almost every issue. More and more software packages are becoming available for various tasks, such as project cost analysis and control, project scheduling and management, space planning and facility management, computer-aided design and drafting as well as building engineering.

In 1995, the UK government commissioned a study, Construct IT: Bridging the gap. It was intended as an Information Technology strategy for the UK construction industry. The strategy identified two key aims for future use of IT in the construction industry:

1. Integrated project communications framework, supporting closer teamwork.
2. Integrated industry-wide information comprising standard component listings, building performance benchmarks, best practices, etc., to improve and inform construction projects.

These are considered long term goals. In the meantime, specific IT systems are developed to achieve incremental improvement in different stages of the construction process.

1.4.4 IT Usage in Construction Team

Between 1997 and 1999, the Building Centre Trust carried out a research project on the use of IT in construction projects in the UK. The study surveyed 80 projects of different sizes and interviewed 403 professionals. The main findings of this study can be summarised as follows:

- Larger organisations invest more in IT compared with smaller organisations. However, the prevailing attitude towards IT development is 'let's wait until others have tried it and we'll follow'.
- The majority of individual professionals have access to a computer (86%), e-mail (97%) and the Internet (88%). However, the use of e-commerce is still limited, only 13% of main contractors had placed an order on-line or by e-mail.
- IT infrastructure is on the agenda of most project teams from the outset, with 15% of the projects having implemented a dedicated project network.
- CAD is the most used specialist software, by 69% of all the professionals, and 99% of the architects.
- Electronic tools were used extensively by individuals in preparing information to fulfil their roles on projects.
- Paper is still the main medium for exchanging project information. Although 79% of product specifications were produced electronically, 91% were distributed on paper.
- The main benefit gained from the use of IT tools on projects is speed of working.

The construction industry may still lag behind some other services and manufacturing industries, such as banking and aerospace, in adopting IT. However, the above survey and other similar surveys have shown that the usage of IT in construction is growing steadily.

1.4.5 IT Construction Best Practice Programme

The IT Construction Best Practice Programme is an initiative to promote the use of IT in the UK construction industry. It is funded by the Department of Trade and Industry as part of the Construction Best Practice Programme. The aim of the programme is to identify examples of successful use of IT in construction practice, publicise them and encourage take-up by more companies so as to improve the business and management practices for the whole industry. It seeks to achieve its aim through the following measures.

- *Case studies*: A series of best practice case studies has been compiled from real companies and real use of IT systems in practice. For each case study, a report is produced, which outlines the business background where an IT solution is implemented, a description of the approach of the IT implementation and the business benefits. These reports are freely available to other companies. The Programme encourages learning from the success across the whole industry.
- *Self-assessment tool*: It is recognise that it is important for a company to establish its IT needs. For this purpose, the Programme provides a self-assessment tool. The tool will help companies to assess where and how IT can contribute to their business success, and measure their progress.
- *'How-to' guides*: The Programme produced a series of practical 'how-to' guides to help companies to adopt new techniques and practices with confidence. These guides provide advice on selection, implementation and management of specific IT systems. Example guides include 'How to manage an IT project', 'A beginners guide to e-business in construction' and 'How to get started in virtual reality'. Again, these guides are freely available.
- *Seminars and company visits*: The programme also organises seminars, workshops and company visits in different regions of the country. These events help to generate awareness and promote knowledge sharing.

The IT Best Practice Programme maintains a website (www.itcbp.org.uk). It is a one stop source of information and guidance on the benefits of effective IT use for the UK construction industry.

1.5 OVERVIEW OF CONSTRUCTION APPLICATIONS

The 'Construct IT Bridging the Gap' report (Department of Environment, 1995) examined the current use of IT in the construction industry and concluded that 'software applications are available to support most aspects of a construction project. They have been designed largely as solutions to specific problems. Applications are particularly strong in design and analysis.' This conclusion is supported by the long list of construction IT applications compiled by CICA. Its Software Directory listed some 1650 programs from over 500 software houses for use in the design, construction and maintenance of buildings (CICA, 2003). For the purpose of this book, we group these programs into six categories:

1. Business and Information Management
2. Computer Aided Design and Visualisation
3. Building Engineering Applications
4. Computer Aided Cost Estimating
5. Planning, Scheduling, Site Management
6. Computer Aided Facilities Management.

Figure 1.4 is a 'roadmap' showing where and when these applications are being used during the construction process. It is not meant to be a definitive picture. The main purpose of the diagram is to indicate the main application areas for the existing discrete software packages in the construction supply chain.

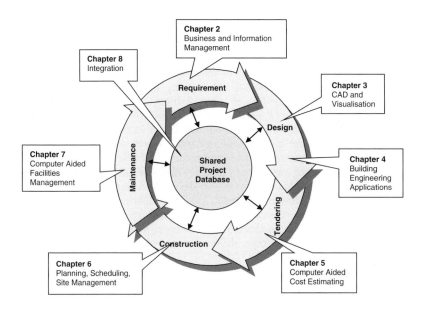

Figure 1.4 Construction process and IT applications

The following is a brief introduction to these applications. Each type will be discussed in more detail in separate chapters in this book.

1.5.1 Business and Information Management

The construction process is an information intensive one during which a huge amount of information is generated and consumed by all the professionals involved. The common type of information includes site survey, cost analysis, design drawings, specifications, regulations, bill of quantities, project planning, job costing and estimates, etc. The information is presented in a range of paper drawings, documents, correspondence, fax, computer files or e-mails. An Electronic Document Management (EDM) system, such as TDOC or OpenDoc, can create an environment within which disparate forms of information can be linked together, in the context of a project or organisation, to achieve easy access and control.

1.5.2 Computer Aided Design and Visualisation

Computer Aided Design (CAD) software is widely used by design professionals. AutoCAD has the largest share of the CAD market (Howard, 1998). Other popular CAD packages include Microstation, ArchiCAD, MiniCAD, FastCAD, etc. These CAD programs have largely replaced the traditional drawing board at the production information stage. The basic function of CAD packages is that they allow the user to build up drawings by manipulating lines, circles, rectangles and texts interactively on the screen. Some architecture-specific CAD systems, such as AutoCAD Architectural Desktop, even provide graphical libraries of commonly used building elements such as doors, windows, etc. (Richens, 1990).

The initial drawing alone does not show the clear advantages of CAD, since experienced designers can draw on the drawing board equally fast. The real strength of CAD lies in its ability to allow 'editing'. Once a graphic is drawn, functions such as delete, move, copy, rotate, scale, mirror, etc., can be applied to any part of it. Other useful CAD functions are available such as repetition of an element at equal distances along a line, around a circle or on a grid or matrix, extending lines, partial erasures of lines and insertion of fillets. These tasks cannot be easily carried out using paper-based media without restarting from scratch. The easy editing features of CAD systems enable designers to explore more alternative building layouts during design (Kharrufa, 1988). Furthermore, since the drawing can be saved at any stage, the designers are able to keep various versions of the building layout for later study. Once the geometrical information on the building design is stored in a CAD package, different views of the building can easily be produced.

CAD systems, with 2D drafting capacity, can help to improve the productivity of the drawing process; applications with 3D modelling enable designers to investigate the building's internal spatial system and its relationship with the surrounding environment. Visualisation and animation systems, like 3D studio, can produce photo-realistic, static and moving images so that the clients can view the final appearance of the building at the design stage. The emerging Virtual Reality (VR) technology even allows the user to interact with the design model and experience the building in simulated reality settings.

1.5.3 Building Engineering Applications

Today, clients of the construction industry have ever-higher expectations. They want their buildings to look good visually, to be safe structurally, to provide comfortable living environments for their occupants, and to consume less energy in operation, etc. The ever more complex demands on the building design process have given rise to the need for a new approach to building engineering design, based on computer software.

Once a building is constructed, it is very costly to correct any design defects. It is therefore important to simulate accurately the building's performance at the design stage so that problems can be identified and solved. Over the years, a variety of methods and algorithms have been developed to predict building performance in thermal, lighting, acoustic, and structural aspects. Because these simulations involve complex and tedious calculations, it was not feasible to carry them out before computers were used. Alternatives in the form of rules of thumb or simplified procedures were often used. As a result, accuracy was compromised. During the last two decades, a range of building engineering applications have been developed for energy analysis, HVAC design, structural analysis, lighting simulation, etc. The benefits of these applications are that they allow designers to evaluate alternative design solutions in order to reach optimum design. They also help to ensure the design complies with Building Regulations requirements. For example, CADLink from Cymap and the HEVACOMP package both offer a comprehensive range of software options for energy, lighting, and building services design.

1.5.4 Computer Aided Cost Estimating

Controlling costs is one of the most important requirements during a construction project. To achieve this control, contractors and sub-contractors must first perform an accurate cost estimation to establish spending targets. Rigorous project accounting must then be employed to ensure that the actual spending will not exceed the target. Project managers or estimators may use a number of methods for preparing cost estimates for construction projects. The simplest, but least accurate, method is an 'educated guess' where the expert comes up with a figure based on previous experience. Today, there are sophisticated computer software packages, such as Esti-Mate, Manifest and FBS-Estimator, which allow project managers to perform estimation and to keep track of project spending (Construction Computing, February 1997).

Computer programs can also assist in the quantity take-off by helping estimators to measure, count, compute, and tabulate quantities, lengths, areas, volumes, and so forth, of objects found in plans and specifications. One way would be to read drawings conventionally and type the data into a spreadsheet or custom take-off program. Some programs, i.e., ConQuest, used in conjunction with a digitising tablet, can read data from paper based drawings, even directly link to CAD drawing files.

Most cost estimating programs can be integrated with databases of costs for labour, materials and equipment. The advantage is that cost data do not need to be re-entered thus improving the speed of estimating and avoiding errors. Computer-based estimating programs deal increasingly well with the data collection, computational, and clerical aspects of estimating. They archive and retrieve large volumes of resource, cost, and productivity information, perform calculations quickly and more accurately, and present results in an organised, neat, and consistent manner. All these virtues are of tremendous help in the high-pressure environment in which most construction estimators often find themselves.

1.5.5 Planning, Scheduling and Site Management

It is a common misconception that computers are of little help on a building site because on-site operation is mainly physical work. In fact, construction work requires careful planning and skilful management of human and physical resources. Computer systems can assist on-site managers to plan ahead, evaluate different options, adopt and execute the most efficient construction operation.

Apart from the wide spread use of planning packages such as Microsoft Project, Primavera, Power Project, etc., to programme and schedule the detailed construction activities, some applications, like JobMaster and ICON, are designed to log and track internal processes during the construction phase, including:

- drawing receipts and distribution
- subcontractor procurement
- issue of instructions
- change control
- outstanding works and defects.

Site operation simulation programs can emulate what happens at a construction site in the real world by representing the workers, machines and materials, and computing the cycle times of each step, taking many uncertain factors into consideration (Paulson, 1995). Their aim is to identify production bottlenecks, reduce under-utilised resources and develop a more productive operation sequence.

Site monitoring systems can address the industry's requirements for increased reporting, improved security and provide accurate and comprehensive information and track movement on site. An example is the Construction Site Monitor (CSM) developed by Public Access Terminals (PAT). CSM is able to instantly produce low cost paper ID badges using standard office equipment by connecting a camera to a computer. Connected to barcode readers, the software is able to log attendance and grant access through turnstiles or road barriers, and print detailed and summary reports for evacuation purposes or financial control.

1.5.6 Computer Aided Facilities Management

Facilities Management (FM) is a relatively new discipline that emerged in the early 1980s. It reflects the wide recognition of the importance of the building operation and maintenance and the impact they have on the life cycle cost of a building. The software available for facilities management has developed from a combination of CAD and database management systems. CAD is often used to present data on departments and locations of individuals together with their services. Special routines enable blocking and stacking studies to be carried out to explore alternative layouts or to reflect organisational change. Typically these show departments as blocks of colour and allow them to be shifted on floor plans and between floors to obtain the best fit.

The database is the most important part of FM software; it holds data on people and their services (telephone extension numbers for example) so that, when they move, their services can go with them. Cable management is often a related function that manages the connection of voice and data cables. The corporate database also holds information on furniture and other assets so that an asset report can be prepared quickly at the end of a financial year.

1.5.7 Integration

Historically, different IT applications are developed by different vendors. They use their own data formats, which are not compatible with each other. As a result, data cannot be exchanged electronically between them. In recent years, there is an increasing awareness of the need for integrated construction processes and a number of research projects are investigating related issues world-wide, i.e. in Finland (Björk, 1987), in the United States (Brambley, 1988), in the European Community (Augenbroe, 1988), and in the UK (Construct IT initiative).

During the last two decades, advances in object oriented programming, database systems and product data modelling technologies have provided a solid platform for integration. Data standards are being developed first by the International Standards Organisation (STEP), and then by the International Alliance for Interoperability (IFC). At present, these standards are still evolving. An integrated project database covering the whole life cycle of construction projects remains a future prospect.

1.5.8 Other Types of Applications

In addition to the above described, there are many general productivity tools used in modern offices. Office automation is a term often used to describe the use of computer systems for typical office activities such as typing, filing and communications. Gone are the days when a secretary had to type every report manually using a typewriter. Using word processing packages, anyone can prepare their reports on computer, making as many revisions as necessary. Desk Top Publishing systems make it easy for anyone to produce output of publishing quality. Documents can be passed between colleagues via networks, or as files on disks, when they work on the same report. The elimination of the need for re-typing has helped to increase office productivity. Presentation software allows every user to make effective and exciting presentation material.

In recent years, computer networking has brought new dimensions to the computing revolution. It enables people to share resources, e.g., printers, software, and file storage within one organisation, and to communicate and exchange information electronically over long distances. Electronic mail, or e-mail is an application that sends messages from one computer to another through the network. It is used by millions of people around the world. It is much faster than a conventional mail service, cheaper and more convenient than telephone and fax.

Office tools and e-mail are no doubt important IT applications for the construction industry. However, because help on their use is very readily available from many sources, they will not be covered any further in this book.

1.6 ON-LINE RESOURCES

Today, the World Wide Web (WWW) has become the most important information dissemination vehicle. There is a huge amount of construction IT resource available on-line. In this chapter and all subsequent chapters, we will provide a list of relevant WWW sites.

Construction Innovation

http://www.rethinkingconstruction.org
Rethinking Construction is the banner under which the construction industry, its clients and the government are working together to improve UK construction performance.

http://www.m4i.org.uk
The Movement for Innovation (M4I) is part of the Rethinking Construction initiative. Its aim is to achieve culture change in the construction industry through regional best practice clusters and demonstration projects.

http://www.cbpp.org.uk
The Construction Best Practice Programme provides support to individuals, companies, organisations and supply chains in the construction industry seeking to improve the way they do business.

Construction Research and Professional Organisations

http://www.cibworld.nl
CIB, the International Council for Building, is an association whose objectives are to stimulate and facilitate international cooperation and information exchange between governmental research institutes in the building and construction sector, with an emphasis on those institutes engaged in technical fields of research.

http://www.riba.org
The Royal Institute of British Architects is one of the most influential architectural institutions in the world.

http://www.rics.org
The Royal Institution of Chartered Surveyors is the home of property professionalism worldwide. This is the information site for property professionals and users of their services.

http://www.constructionconfederation.co.uk
The Construction Confederation is the leading representative body for contractors in the UK, representing some 5,000 companies who are responsible for over 75% of the industry's turnover.

http://www.biat.org.uk
British Institute of Architectural Technologists is recognised as the leading body in Architectural Technology qualifying Architectural Technologists in the UK.

http://www.crisp-uk.org.uk
The Construction Research and Innovation Strategy Panel brings together Government, clients, industry and the research community to consider research priorities for construction.

http://www.ciob.org.uk
The Chartered Institute of Building is the leading professional body for managers in construction.

http://www.construction.co.uk
Co-construct is an umbrella body for five professional and research associations. Its aim is to promote best practice in the construction industry.

http://www.ciria.org.uk
The Construction Industry Research and Information Association is a UK-based research association concerned with improving the performance of all involved in construction and the environment.

Construction IT

http://www.itcbp.org.uk
The IT Construction Best Practice Programme is a one-stop source of information and guidance on the benefits of effective IT use for the UK construction industry.

http://www.cica.org.uk
The Construction Industry Computing Association is dedicated to the effective use of information technology in the construction industry. It maintains a software directory of construction IT applications.

http://www.construct-it.org.uk
Construct IT For Business has been set up to coordinate and promote innovation and research in IT in Construction in the UK to improve competitive performance of the UK construction industry and to act as a catalyst for academic and industrial collaboration.

http://construction-institute.org/structur/cii132.html
The Information and Communication Technologies for the construction industry web site has a collection of resources for construction IT applications.

http://www.connet.org
The European Construction Information Service Network is an Internet gateway to information and advice on the built environment in Europe.

http://www.cite.org.uk
Construction Industry Trading Electronically (CITE) is a collaborative electronic information exchange initiative for the UK construction industry.

Construction Information

http://www.constructionplus.co.uk
Construction Plus is the Internet arm of the Emap Construction Network, the largest information supplier to the UK construction industry. It provides news and information about construction industry and IT applications.

http://www.aecinfo.com
On-line construction resources. This provides fast and easy access to over 300,000 pages of building material information including specifications and CAD details.

SUMMARY

The convergence of computing and communication technology offers unprecedented opportunities to improve business efficiency and performance. Although the construction industry has been slow in adopting computing technology compared with some other manufacturing and services industries, a large number of software programs have been developed during the last two decades. This chapter briefly reviewed the main types of construction IT applications. Detailed discussions on their main functionality and business benefits will be given in the following chapters.

DISCUSSION QUESTIONS

1. Think of several examples in your daily life where computers are used.
2. What are the main features that make computer systems potentially useful to construction professionals?
3. Discuss how each type of IT application can change the way construction professionals work in practice.

CHAPTER 2

Business and Information Management

LEARNING OBJECTIVES

1. Understand the principles of management.
2. Distinguish between consultants' and contractors' applications.
3. Identify applications for resource analysis and planning.
4. Anticipate new developments in business management.
5. Recognise the need for valuing and accessing knowledge.

INTRODUCTION

'The construction industry is as much a manager of information as it is of materials', was said in 1990 by John Hollingworth, then head of IT at Wimpey construction, in his contribution to Building IT 2000 (Building Centre Trust, 1990). That statement has become even truer since then with the arrival of the Internet and computers on the desks of most employees based in offices. The construction industry manages large teams of people based in different companies to design and construct complex buildings and other structures.

Businesses of all types have to review their processes in the light of new technology and to maintain their competitiveness, and construction is no exception. It has had a very traditional structure, with clearly defined roles for each consultant and each trade. Greater integration of project teams and tools for sharing information are leading to these roles being questioned and more efficient processes being tried out. Business Process Reengineering is a management technique for doing this. Integration of separate software tools is leading to Enterprise Resource Planning systems that can hold most of the information a company needs for managing its processes. What cannot be held in computer networks, the knowledge that depends upon people for its application, is also the subject for new Knowledge Management systems that record where the knowledge can be found.

Information management has become a serious problem with the growth of means of delivering it. There is now too much and it is of variable quality. How do we find what we need and establish its reliability? Techniques such as data warehousing and data mining are needed to find what is required, and intelligent search engines and agents can help to identify the information we really want. Most businesses are now dependent upon IT for their survival and the IT systems themselves must be managed and their value to the company properly evaluated. The future of business and information management will be crucial to the effectiveness of the construction industry. Information is power.

2.1 FUNDAMENTALS OF MANAGEMENT AND INFORMATION

2.1.1 Management Theory

Management is, on the one hand, common sense application of control that can be carried out, for example, by small builders based on their experience, in an unconscious and often very effective manner. On the other hand, as soon as a business becomes too large to allow one person to communicate regularly with all the employees, say when staff numbers exceed 10 or 20 people, then the more complex processes of management science start to apply. There are three main schools of management theory:

Table 2.1 Three main schools of management theory (after Laudon, 2000)

Technical-rational	The precision with which a task can be done, the organisation of tasks into jobs, and jobs into production systems
Behavioural	How well the organisation can adapt to its external and internal environment
Cognitive	How well the organisation learns and applies know-how and knowledge, and how well managers provide meaning to new situations

All these are relevant in construction but the main management effort has traditionally been put into technical-rational organisation of the many tasks to be done. Now, with people as the main asset of most companies, there is more attention given to behavioural management of people, and the growing recognition of the value of knowledge is leading to cognitive techniques being applied.

2.1.2 Data, Information and Knowledge

The term information is often used loosely to cover data, information and knowledge, and it is necessary to distinguish between these.

- *Data* is (or are) collections of facts, measurements or statistics.
- *Information* is organised or processed data that is timely and accurate.
- *Knowledge* is information that is contextual, relevant and actionable.

Data is what we tend to have too much of and, although necessary for management or for operating any computer system, it needs to be carefully selected for its timeliness and accuracy, thus becoming *information*. *Knowledge* is what is most useful and in shortest supply. It has strong experiential and reflective elements that distinguish it from information. An example is the timesheet data that all consultant offices collect. This tends to be categorised according to which projects each employee is meant to be working on. The reality may be rather different. The information that is needed by the company is how much they can charge to their clients. Hidden within this is the real knowledge that few companies ever discover, which is how much time has really been spent and how profitable is each project.

2.1.3 Management Information Systems

The main role of management is to make decisions and, although traditionally this would be done without needing to be explained, in large organisations there is a need to justify actions and often take decisions collectively. For this reason a number of types of Management Information System (MIS) have been developed.

Table 2.2 Traditional and contemporary management (after Laudon, 2000)

Management schools and tasks	Traditional	More recent	Role of IT
Technical-rational			
Analysis of work	Time/motion studies	Analyse teams or whole processes	Business process reengineering
Administration	Hierarchical reporting	Analyse information flows	Work-flow software
Behavioural			
Planning	Top-down centralised	Decentralised involving all staff	Use of Internet to involve more staff
Organising	Stable division of labour	Self-organised project teams	Networking of teams
Leading	Inspiration or threat	Enable staff to use their best abilities	Networks for supervision
Controlling	Precise control systems	Control at team level	Real-time organisation
Innovating	Centralised R&D	Ideas invited from all staff	IT conferencing, groupware
Environments	Hostile and competitive	Proactive with potential for alliances	Monitor changes
Cognitive			
Sense-making	Imposed by managers	Informal group activities	Systems which cannot be biased
Organisational learning	Captured by routine procedures	Captured by information systems	Create, store and disseminate knowledge
Knowledge base	Financial and physical assets	Core competencies and knowledge	Knowledge management

2.1.4 The Latham and Egan reports

In the construction industry the design process is one of constant innovation and yet the documentation, approval and construction processes that follow are relatively unchanging. The recent examinations of the whole construction process in the Latham (1994) and Egan (1998) reports have applied management thinking from other industries to construction. Both have indicated that construction costs could be reduced substantially and that productivity in the industry has not increased greatly in spite of information technology. The 'productivity paradox', states that in spite of wide deployment of information systems, no industry has yet been able to prove that its overall productivity, that is the value of turnover per person day worked, has improved. It is hard to believe this for the banking and insurance industries, where systems have replaced people on a large scale, but it is not so hard to see how it applies to construction.

In the UK these reports are leading to increased use of a number of management techniques and process changes:

- *Supply chain management* in which electronic communications and long term relationships help to ensure integration of processes carried out by different firms.
- *Public-private partnerships* in which private money is used to finance facilities for public use through the health or educational services.
- *Build-Own-Operate-Transfer* (BOOT) in which contractors or consortia tender for providing and operating a facility for many years.
- *Benchmarking* involves the development of key performance indicators for various measures of performance by which companies can continuously improve.
- *Partnering* in which whole project teams are appointed for a series of projects so that they can establish efficient working methods and even share risk and profit.

2.1.5 Professional Office Management

The typical consultant office management system, in contrast to that used by contractors, takes timesheet data, allocates it to projects, combines it with chargeable expenses and generates invoices to clients. This data can then be used in combination with information about future prospects of work, to provide knowledge and plan the resources needed by the company. Will it need to take on staff or reduce in size? Will it need to acquire different skills or new equipment? It is notoriously difficult to plan for the fluctuating demands of the construction industry. This is subject to changes in the national economy since capital spending is the first to suffer in a downturn. Each project depends upon the client's resources and changing needs, and planning permission often takes time to obtain. It was noticeable, during the serious recession around 1990, that companies selling office management software showed high levels of growth.

Table 2.3 Decision making and information requirements

Decision stage	Information needed	Examples of software systems
Intelligence	Timesheet data	Database
Planning	Current & future work	Resource analysis
Choice of options	What-if simulation	Spreadsheets, decision support
Implementation	Graphics, charts	Spreadsheets, presentation systems

Most small offices can manage their resources entirely using spreadsheet programs, such as Microsoft Excel. The charting facilities allow presentation of information visually, which is much easier for a partners' meeting to assimilate quickly. If multiple projects are planned in detail and the timing of tasks is critical, then a network analysis system, such as Microsoft Project, would be useful. There are a number of specialist professional office management systems linked to accounts on the UK market and the following are a few of those listed in the CICA 2003 Software Directory (CICA, 2003).

Buttress	Definitive Computing	www.dclsoftware.co.uk
Progression AEC	Ramesys	www.ramesys.com
Shortlands	Datapro	www.datapro.co.uk
Officebase	Tech Computer Office	www.tcosoftware.com
Cost & Works	Causeway Technologies	www.causeway-tech.com
My SAP HR	SAP (UK) Ltd	www.sap.com/uk

For larger organisations, and to integrate all the different modules available for management, Enterprise Resource Planning (ERP) systems may be necessary. These were developed in manufacturing industry to merge financial control systems with material handling, order and production control. They are therefore more appropriate to contractors and materials suppliers than to professional offices. However the main suppliers of these, SAP and BAAN, have developed optional modules for service-based companies. The three main characteristics of ERP technology are scope, configurability and integrativeness. A system must address the needs of the company without being too heavy a burden on its administration. Each company will have different requirements and suitable combinations of modules will need to be configured to meet these. Integration is the main objective of using ERP rather than a series of isolated applications, and this implies using a single database and good communications to allow access from all parts of the organisation.

Two optional modules are of particular relevance to professional offices in construction: Customer Relations Management (CRM) and Document Management, covered later. Starting from an office maintaining its Christmas card list on a word processor, CRM systems have become large databases with details of clients' preferences and past commissions and can be accessed on-line when making contact with potential customers.

2.2 BUSINESS PROCESS REENGINEERING

2.2.1 The Five Forces Model

Management science has been greatly influenced by the work expressed in Porter's five forces model for an enterprise to position itself in relation to others (Porter, 1980).

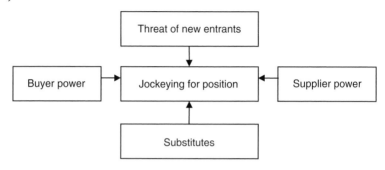

Figure 2.1 Porter's five forces model (after Porter, 1980)

This work was based upon studies of manufacturing industry and there are different factors applying in construction. It is hard to see substitutions of essential types of building, for example, but buyers have all the power in deciding when and what to build. Jockeying for position was traditionally done by consultants on golf courses and at social gatherings. This has changed with consultants having to tender on their fees just as contractors have always had to tender on construction price.

In future, in the European Community, it is possible to see threats from new entrants from the countries of Eastern Europe, where labour costs can be one tenth of those in the west. Currently recognition of qualifications and trade agreements limit this, but they may have less effect on movement of labour in future. Heavy materials are costly to transport over large distances but information, consisting of bits rather than atoms, as memorably compared by Nicholas Negroponte (1995) in 'Being digital', can be transported anywhere. Much commercial software, for example, is now written in India to take advantage of cheap, skilled labour. What is to prevent design work being carried out in a different part of the world, except the cultural variations that are still reflected in local building traditions.

2.2.2 Changing Requirements and Business Strategies

IT also leads to the need to consider changes in future building requirements. The demise of the city centre office building has long been predicted as a result of technology allowing some people to work from any location, even their homes. Are large university complexes needed with the potential of distance learning? Even if these changes come about, there will be new building types needed – more cultural buildings in which to spend our leisure time, call centres to handle e-business and large secure computer centres to hold all the electronic data.

Porter has also identified four approaches to business strategy to help companies to focus on the areas in which the five forces model indicates the prospect of greatest success. It is important to have a clear business focus before implementing the IT systems that may be necessary to achieve it.

Table 2.4 Four generic strategies (after Porter, 1980)

	Lower cost	Differentiation
Broad target	Cost leadership	Differentiation
Narrow target	Cost focus	Differentiation focus

With competition for work applying in almost all areas of building design and construction, it becomes necessary for an organisation to decide whether it is going to compete on cost and, if so, on a broad or narrow target market, or whether it will go for specialisation and quality. Examples of such different approaches are: the volume house builders aiming at cost focus, or a multi-discipline consultancy, such as the Building Design Partnership, offering differentiation by combining all the consultants disciplines in one firm. These appraisals of a business that any company or partnership feels obliged to make can lead on to the process of reengineering the business. This is a sequence of evolutionary and revolutionary changes, influenced by IT, which drive organisational change.

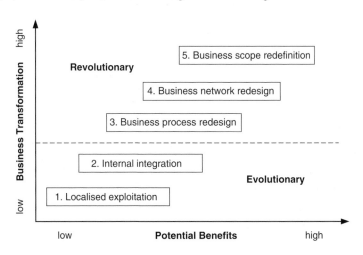

Figure 2.2 Business process reengineering

2.2.3 Business Transformation

Most companies start from localised exploitation of IT by applying it to individual systems typically, in construction: word processing, accounting, project management and draughting. Then individual systems become integrated within the organisation. Up to this stage there may need to be little change to the strategic processes in the company and certain productivity gains should be achieved. These can be measured simply by analysing person and machine time used to produce documents, both before and after introducing IT systems.

The next stages are the ones that can transform a business or the processes it uses, and involve greater integration with customers and partners. The ways in which buildings are designed or constructed need to be rethought. Relationships with clients should become long-term ones, and new areas of business, such as property development, might be considered. The result of this could be a revolutionary change. Many construction organisations have had to go through this process due to the ups and downs of construction workload. One example was the swapping of Wimpey's construction business for Tarmac's housing division a few years ago.

2.2.4 An Example of Partnering

Other changes have come from organisations commissioning a large volume of building work, of which BAA is the most quoted example in the UK. Being led at the time by Sir John Egan, who had come from the motor industry, BAA felt that it was not getting value for money for all the airport buildings it constructs and maintains. It set up framework agreements with consultants, contractors and suppliers for five-year periods, even though this involved lengthy selection processes. This length of collaboration, now extended to 10 years, allows project teams to set up working methods that might be unproductive over the duration of a single contract. It provides an opportunity to introduce new technologies such as, in the case of BAA, the Industry Foundation Classes for shared building models.

Figure 2.3 Heathrow Terminal 5 model (reproduced with the kind permission of BAA)

2.3 KNOWLEDGE MANAGEMENT

This term Knowledge Management (KM) is relatively new to the construction industry. It indicates a recognition of the intellectual property rights that any organisation owns. Most of these are locked up in the experience of many years of work and completion of many projects, and they mainly reside in the heads of the staff. For this knowledge to be used for organisational success, it should be recognised as a form of capital, and must be capable of being exchanged between people and of growing. Knowledge takes two basic forms: tacit and explicit. Explicit knowledge can be captured and stored and used without reference to others. Tacit knowledge has been described as a cumulative store of experiences, mental maps, insights, acumen, expertise, know-how, trade secrets, skill sets, understanding and learning that an organisation has. It is difficult to codify and requires effective human organisation and some explicit knowledge to apply it constructively.

Since the tacit knowledge resides in peoples' minds, there should be ways of recording who has what experience, on past projects for example, and ways of getting these people together with others who need that knowledge, such as in focus groups. Computer systems have only a peripheral role to play in this by offering databases of projects and the people who worked on them, or CVs for staff that indicate their special knowledge. Some of this may be jealously guarded but most people are flattered to be asked about their fields of expertise.

2.3.1 The Conversion of Knowledge

Tacit knowledge can be articulated through observations even though those who have tacit knowledge are not able to express it. To make use of tacit knowledge for competitive advantage, it needs to be articulated and utilised by companies and their partners (Nonaka, 2000). Tacit knowledge needs to be converted into explicit knowledge in order for it to be shared and utilised by others. Nonaka (1995, 2000) described the conversion of tacit to explicit, and vice versa, by categorising it into four type of conversions namely socialisation, externalisation, combination and internalisation (Figure 2.4).

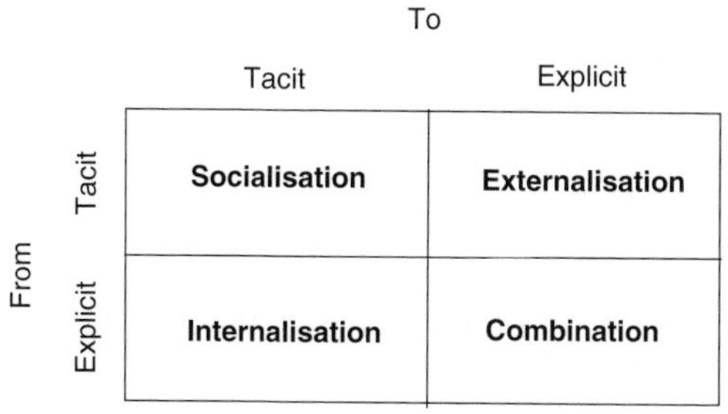

		To	
		Tacit	Explicit
From	Tacit	Socialisation	Externalisation
	Explicit	Internalisation	Combination

Figure 2.4 Knowledge conversion process (after Nonaka, 2000)

Socialisation is the process of sharing tacit knowledge between individuals. It is experiential, active and a 'living thing', involving capturing knowledge through direct interaction between people. This process depends on having shared experience, and results in acquired skills and common mental models. The process for making tacit knowledge explicit is *Externalisation*. There are two aspects in this process. One is the articulation of one's own tacit knowledge – ideas or images in words, metaphors, and analogies. The other is eliciting and translating the tacit knowledge of others into a readily understandable form, e.g., explicit knowledge. Dialogue is an important means for both. During such face-to-face communication people share beliefs and learn how to better articulate their thinking, through instantaneous feedback and the simultaneous exchange of ideas. The conversion process between different forms of explicit knowledge is called *Combination*. This is the area where information technology is most helpful, because explicit knowledge can be conveyed in documents, e-mail, databases, as well as through meetings and briefings. The process usually involves collecting relevant internal and external knowledge, disseminating, and editing/processing to make it more usable. *Internalisation* is a process of experiencing knowledge through an explicit source. The newly created explicit knowledge is converted into tacit knowledge of individuals. Practical training, exercises and learning by doing are important to embody explicit knowledge.

2.3.2 Process of Knowledge Management

Knowledge management has two aspects, (1) managing knowledge itself, typically using some IT technology, and (2) managing people who create knowledge and encouraging a knowledge sharing culture within an organisation. The process of knowledge management can be viewed from three main perspectives. These include:

- *IT perspective*: This is concerned with the tools and techniques for capturing and distributing information and knowledge (data warehousing, Case Based Reasoning, GroupWare, Intranets, etc.). Knowledge Management is not an IT process but IT does, and will, play an important role in facilitating knowledge management.
- *Social perspective*: Knowledge management also has a significant social and political dimension. Aspects such as physical proximity, knowledge (management) domains, and ownership, etc., are all social issues that need to be addressed and taken into account in any knowledge management system.
- *Cognitive perspective*: This pulls the other two perspectives together and is concerned with the processes by which individuals or groups can not only share knowledge but also use it in problem solving to create new knowledge. It involves the sharing and learning of tacit knowledge.

The complete knowledge management cycle involves a number of stages:

- Creation of knowledge
- Capturing knowledge
- Refining knowledge
- Storing knowledge
- Managing knowledge
- Disseminating knowledge.

2.3.3 Knowledge Based Systems

Knowledge Based Systems (KBS) provide one technique for encapsulating such subjective information. These were originally conceived as having intelligence in interpreting human knowledge and applying it to defined problems, indicating the probability of different solutions and explaining their reasoning in natural language. The few systems that have been successful in construction might not have been classified as knowledge-based by these criteria. Elsie was a system developed for the RICS at Salford University in the 1980s and used the combined expertise of senior quantity surveyors to build cost models usable at the early design stage. ICI, as one enthusiastic user of the system for office buildings, found that a guide price could be obtained with it more quickly and cheaply than by commissioning feasibility studies from architects and then going through the normal quantity surveying process. It is doubtful if it was a true expert system by the criteria listed above, but it provided useful guidance for standard building types on unconstrained sites.

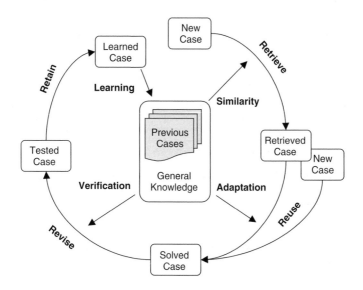

Figure 2.5 Knowledge management using Case Based Reasoning (after Bush, 1999)

Case Based Reasoning (CBR) is a technique widely used to develop KBS systems. It is a method of solving new problems by adapting solutions that were used to solve old problems. Its principle is based on the assumption that similar problems have similar solutions. Figure 2.5 shows the problem solving process using CBR, which involves a four-stage cycle:

1. *Retrieve* the most similar case(s) from stored previous cases.
2. *Reuse* the retrieved case(s) to attempt to solve the problem.
3. *Revise* the proposed solution if necessary.
4. *Retain* the new solution as a part of a new case.

2.3.4 The Role of IT in KM

Table 2.5 and the following explanations identify the areas where the impact of information technology might have a positive influence on knowledge management applied to the engineering design process (Bush, 1999).

Table 2.5 The impact of information technology on knowledge management
(after Davenport, 1993)

Impact	Explanation
Automating	Eliminating human labour from the process
Informational	Capturing process information for purposes of understanding. Delivering information to the right person at the right time.
Sequential	Changing process sequence, or enabling parallelism
Tracking	Closely monitoring process status and objects
Analytical	Improving analysis of information and decision making
Geographical	Delivering information and knowledge across distances, facilitating collaboration across distances and co-ordinating processes across distances
Integrative	Co-ordination between tasks and processes
Intellectual	Capturing and distributing intellectual assets

Automating: The most commonly recognised benefit of IT is its ability to eliminate human labour and produce a more structured process. A Knowledge Based System (KBS) delivers knowledge in a structured and logical form dictated by rules, case-based reasoning or neural networks. They can eliminate a vast amount of time consuming, mundane and tedious work. (For example London Underground controls passenger access to platforms using a KBS that monitors video images of platforms showing the state of congestion.) As well as reducing

---

Case Study
Xerox benefits from a Knowledge Based System

An example of KM can benefit a company and its customers, is given by Xerox Corporation.

Xerox developed a knowledge management system called Eureka for copier service technicians to share best practices and solutions to problems. Its effectiveness was demonstrated when a leading copier developed intermittent failures in different parts of the world. It was unable to identify the problem and had to replace six machines. It occurred again in Rio de Janeiro, where replacement was estimated to cost $40,000. A service technician in Montreal had traced the problem to a 50 cent fuse holder that tended to oxidise and had to be swabbed with alcohol occasionally. He had posted this tip on Eureka and, when this came online in Brazil in Portuguese, the problem was solved. Personal recognition motivates technicians to submit tips to Eureka since each tip has the author's name attached. By early 2000 Eureka contained 5000 tips, and was available to 44,000 technicians worldwide.

Usually most of the knowledge of this sort never gets into a system that is accessible to others. The problem that managers have to solve is how to extract such knowledge from the people who hold it, and what are the chances of its ever being reused. Most companies concentrate on linking people so that, when a problem arises, the person most likely to have the solution can be contacted. However people move to other companies or retire, and the dilemma is then how much of their experience to try to capture. Sometimes companies which are trying to change their culture want out of date experience to be lost, so the knowledge which is captured needs to be refined to relate it to the current needs of the company.

---

the cycle time, this provides greater opportunity for the engineers involved to spend more time on innovative technical considerations.

Informational: Capturing process information for future analysis and optimisation is a key element of many knowledge management systems. Most KBS are also concerned with the delivery of the right information to the right person at the right time.

Sequential: Although there is usually a natural hierarchy between concepts, objects or frames of a KBS, it is possible for systems to accommodate concurrent engineering practices by allowing the refinement of one concept whilst investigating another. There are also instances where KBS are used to analyse workflow or process controls and advice on sequential or parallel work practices.

Tracking: Monitoring and tracking process status or objects can have a significant impact on productivity and avoid bottlenecks. KBS have successfully been employed in this area by Mitsubishi Electric Corp., Nestle, and Daimler-Benz to name a few.

Analytical: KBS can bring to bear an array of sophisticated analytical resources that can add value to the product or provide information and understanding that otherwise would be missing or difficult to acquire. Again the effect here is to reduce project cycle time and/or improve understanding of technical issues.

Geographical: To employ our intellectual capital it is essential that geography no longer acts as a barrier or restriction. IT communication technologies applied to knowledge management now make this possible.

Integrative: By alerting experts to information and knowledge locations and providing tailor-made, easy-to-use applications, specifically suited to the processes involved for a project it is possible to ensure that only one team carries out all aspects of the project processes. Small, multi-disciplined teams are at the heart of concurrent engineering initiatives and are thought to be responsible for dramatic productivity improvements.

Intellectual: Knowledge management is all about distributing and accessing a company's intellectual capital.

2.4 INFORMATION MANAGEMENT

2.4.1 New Techniques for Managing Information

With most data available electronically and that on the Internet conforming to no structure except that of the HTML language of the World Wide Web, businesses need help to find the information that is useful to them. Search engines are the response to the anarchy of the Web. These trawl around the web and extract and classify information in ways that are more useful to people. A keyword search through Google or Altavista can then locate sites from its more structured data very rapidly. The results may still be rather random. An example of this was at one of the first Internet seminars given by CICA in the early days of the Web. There were few sites at that time and those attending the seminar were rather surprised to see there were 10,000 hits when searching on the word 'construction'. Further investigation showed that these were almost all sites 'under construction' and had nothing to do with the industry.

Intelligent agents are one response to the chaos of electronic data. These software tools can learn, or be taught, how users work and provide assistance in their daily tasks. They can help to eliminate junk mail, select types of information that users want, and customise a software system to suit the user's preferences. They can also work for the information providers in placing 'cookies' on a user's computer disc, and these small programs feed back intelligence about the user's habits. Anyone working on line to the world will have left a trail of information that can be very valuable to marketing people.

Within more structured sets of data such as company databases, techniques such as data mining can be used. These apply artificial intelligence methods to collect information. An example is the provision of store cards by supermarket chains. Most customers think they are only to breed loyalty and give small discounts. In fact they link the buyer's details to all the products they buy, and it is possible to derive valuable marketing information from this and make targeted special offers. In construction valuable information can be mined from logs of Project Webs. These are Extranets shared by project teams and through which most documents pass. These logs can be analysed to see which members of the team are uploading most documents and which are downloading them. Some documents are well used and are obviously the most important ones, while others may never be accessed by anyone, and may be quite unnecessary. As more Project Webs are used, it should be possible to rationalise the production and use of documents.

2.4.2 Electronic Document Management

A major point of discussion in construction at present is whether the concept of a document is fundamental to communication of information. The industry is used to paper documents but, as most of these are now generated electronically and may be based on larger collections of data such as databases or product models, why exchange documents which are just extracts from these. Ideally the fundamental data would be accessible to all, but the concept of a document is one that allows past experience with paper to be transferred into an electronic environment. It will take a long time for reliance on documents to be superseded. Too many construction projects end in the law courts and lawyers are still attached to written documents.

Computer systems to handle documents have become essential to keep track of revisions and record to whom documents have been issued. They are an application of database technology that may be linked to Project Webs or Enterprise Resource Planning systems. These systems store and retrieve structured and unstructured documents electronically. They are known as Electronic Document Management (EDM) systems (Figure 2.6).

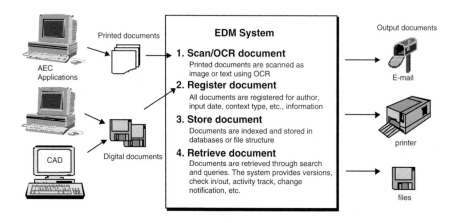

Figure 2.6 Configuration of an electronic document management system

The aim of an EDM system is to create an environment within which disparate forms of information can be linked together in the context of a project or organisation to achieve easy access and control. The EDM environment addresses the following aspects of data management:

- Efficient location and delivery of documentation.
- The ability to manage documents and data regardless of the originating system or form.
- The ability to encompass and integrate with existing computer or paper based systems in the context of a construction project.
- Control of the access, distribution and modification of documents, with the ability to mirror existing company procedures.
- The provision of tools to edit documents and add mark-up information whatever the source of the document.
- The support of both paper-based and digital documentation, including importing of scanned documents.

Document management is available on many stand-alone systems of which Documentum www.documentum.com, Docs Open www.pcdocs.co.uk, and Recall www.kalamazoo.co.uk, are some examples available in the UK. If an ERP system contains details of projects it may be advantageous to link the drawings and other documents related to these, to the rest of the enterprise information. However, many of these documents will need to be accessible to other members of the project teams in other companies, and security considerations normally prevent outsiders from being given access to ERP systems.

2.5 COLLABORATIVE WORKING SYSTEMS

Construction projects always involve the collaboration of a multi-disciplinary project team located in different parts of the country; some may be on-site, others located at an administrative office. Collaborative working using computers is the theme of two research areas, computer mediated communication (CMC) and Computer Supported Co-operative Work (CSCW). In practice the two areas often overlap in producing actual technical solutions. CMC is concerned with both synchronous and asynchronous communication using computer networks as a medium. The communication media include audio, visual or a mixed format. CSCW applies CMC technology to solutions for collaborative working practices by providing a centralised work storage, version control, concurrent work processes functions.

The advent of the Internet has greatly enhanced the operational scope of both CMC and CSCW. There is now a wide range of ready-made tools aimed at supporting projects where participants are potentially widespread. Lotus Notes GroupWare and the Basic Support for Collaborative Work (BSCW) prototype (Bentley, 1997) are good examples of CSCW systems. The following applications are specifically aimed at the construction industry:

BIW Information Channel   www.thebiw.com
Build on-line   www.build-online.com
Business collaborator   www.businesscollaborator.com
Causeway Collaboration   www.causeway-tech.com

2.6 E-BUSINESS

2.6.1 Growth Patterns

The traditional growth pattern for any communicating technology has been the S-curve with any new means of communication, fax for example, growing slowly until the critical level, typically about 30% of a market, and then rapidly accelerating towards saturation level (Howard, 1996). Full saturation of a market is rarely reached because a substitute product or service often appears, such as e-mail taking over from fax. E-business, in its early years, has seen a different pattern of growth. It was so obvious that many information-based businesses would be more efficient trading by electronic means that investment rushed into the first dotcoms in the 1990s at a level that could not be sustained. The result was a crisis of confidence in 1999/2000 that suggested a different growth pattern shown in Figure 2.7 from the Gartner Group (2001). This indicates that, after a period of consolidation, there will be steady and sustained growth.

     The figure also indicates a post-Internet type of business. This might be the result of changes in the Internet as it becomes increasingly corrupted by viruses and unsolicited e-mail. More specialised Extranets would then be needed and these could become virtual storefronts or other forms of simulating the ways in which we are used to shop.

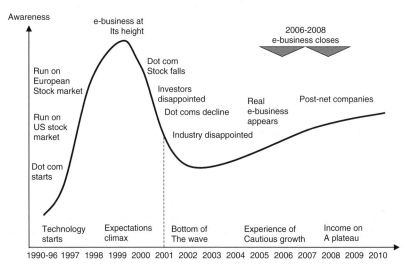

Figure 2.7 The rise and fall of e-business companies (after Gartner Group, 2001)

2.6.2 Types of E-business

A number of different types of e-business model have been tried out with varying degrees of success. One way of testing the success of the examples given below would be to follow the URLs of their websites and see if they are still accessible.

Table 2.6 Internet business models (after Laudon, 2000)

Category	Description	Examples
Virtual storefront	Sells goods or services on-line delivers by traditional means	www.amazon.com www.wine.com
Marketplace concentrator	Data from many providers via one outlet-building products	www.barbourexpert.com
Information brokers	Provide product, pricing and availability information	www.travelocity.com www.railtrack.co.uk
Transaction brokers	Buyers can view rates and terms and complete the deal	www.ameritrade.com
Electronic clearinghouses	Provide auction-like settings for products which change	www.bid.com www.onsale.com
Reverse auctions	Bidding to multiple sellers to buy at a buyer specified price	www.priceline.com
Digital product delivery	Sells and delivers software and other digital products	www.photodisc.com
Content provider	Creates revenue through subscriptions or advertising	www.wallstreetjournal.com www.thetimes.co.uk
On-line service provider	Services and support for hardware and software users	www.microsoft.com

These business models are constantly changing and some will prove of lasting value while others may decline. They all require substantial support and reliable computer networks. They also require secure payment systems and will be increasingly vulnerable to fraud. The computers used for the major services are housed in secure buildings with alternative back-up systems available. At the other end of the scale, anyone can do business from home with a PC connected to the Internet, but there is very little security for the customer in this type of operation.

2.6.3 Portals

With the Internet being so chaotic it can often take a long time to find useful information. One response to this has been to encourage users to go through portals to reach a number of related facilities. Search engines, such as Yahoo structure the URLs they collect so that they can help focus searches. Other successful websites, once they have captured a particular market, can divert their regular users to other areas. This can be risky and Amazon, best known for its book sales, set up links with suppliers of other products. This caused them some problems coping with other companies in a dotcom market that was seeing many failures.

In construction, a major goal is to have portals for access to building product data so that it can be concentrated through a small number of access points, and

---

Case Study
Kajima Adopt BIW for its ASDA Project

Kajima Construction Europe (UK) Limited adopted the use of Information Channel (IC) from BIW Technologies (BIW) during 2000. As one of Asda's partner contractors, Kajima was to carry out a new £10 million superstore development in Wrexham, North Wales, for the retailer. Asda wanted fast track delivery, and Kajima set a very tight target for the design and build.

The Wrexham design phase lead-time was short, and there was enthusiasm from Asda, Kajima and the other suppliers for the concept of using an online system to hold information centrally. BIW's Information Channel assists project collaboration and supply chain integration. It provides a secure project-specific information hub or exchange. Every team member uses this for creating and sharing all project data via the Internet, using a standard web browser, and with no need for any specialist software.

Information Channel was implemented for the Wrexham project to good effect. Because client and supplier team members could simply log in to view drawings, there were useful cost savings of more than £11,000 on printing and £1,000 on postage. The Project Architect reported an 8% increase in efficiency. Traditionally Kajima finds the drawings issue/review/re-issue process can take ten days; here the team was able to reduce this process to an average of just two days. The best time noted was four hours! Around £31,000 was saved on design team fees. The number of meetings decreased as the feeling of control grew. Using other innovative techniques and materials during the build phase, the team took just over 13 weeks to complete the superstore - which Asda was able to open a day ahead of schedule.

Source: BIW Technologies

alternative products can be compared. There is some resistance to this by manufacturers who want to control their customers and not share them with competitors. There is an important role for information brokers such as Barbour Index and the Building Centre in the UK but, if there are too many, the danger is that each will carry a small part of the whole product range. Other types of portal are used by consortia of companies to buy in bulk so that they obtain discounts which individual companies might not be offered.

E-business has the potential to change all sorts of transactions. They can become two-way with everything being negotiable or auctionable. Buyers and sellers can group themselves to obtain greater leverage. Surplus items can be sold off cheaply at the last minute. This should help to create efficiency and reduce prices, but the process of jockeying for position in rather ill-defined markets will continue to be difficult for some time.

E-business is already transforming some areas of commerce. Airlines are a good example where the traditional process of ticketing and discounts to travel agents is starting to break down, for private travel anyway. The low-cost airlines reduce their costs by electronic means and, although they offer telephone reservations, they give significant discounts for booking on line. In construction the changes are only starting to be felt through a few reverse auctions.

2.7 PROCUREMENT AND MANAGEMENT OF IT SYSTEMS

2.7.1 IT System Procurement

In the past companies have often acquired new computer systems without adequately considering their implications. It is easy to download shareware or buy a package just to find out what it does, and to deploy it across a company without considering all the implications. The first requirement is to consider the business need. What advantages might result to the company and its work and staff. Secondly the costs should be estimated and the original price of the software may be the smallest of these. What are the hardware implications, the staff needs and the process changes required, to deploy the system fully?

In the chapter on 'Implementing an IT strategy in practice' in Strategic management of IT in Construction (Betts, 2000), Derek Blundell summarises the issues thus:

- Planning, selecting software and solutions, rolling out systems and supporting them.
- Implementation strategy, process change, setting up a project team, seeking integration, sequencing, defining timescales and resources and managing applications.
- Specifying systems, choosing between bespoke systems and packages, selecting suppliers or software development processes.
- Auditing skills, preparing data, training and user consultation.
- Help desks and user groups.

The stages in the process should include: planning, software selection, implementation and support. The starting point is business strategy and there may be a need to consider business process change in order to get full benefit from new IT systems. A User Requirement Specification should be developed with the help of those necessary to implement the systems. If there are packages that already meet the specification, or could be adapted to do so, then a number of these should be evaluated. If not it may be necessary to carry out, or commission, software development. This is a route that needs to be planned very carefully. There are many examples of systems over-specified which take too long to produce or are obsolete before they reach production. Specifications should be hierarchical with requirements categorised as:

- Essential – the system must have these features or it cannot be used
- Beneficial – the system can work without them but they will add value
- Nice to have – only if they can be provided easily.

The timescale of the introduction should also be planned and, if possible, phased. For example a new system should be tried out on one group first and their reaction tested before deploying the system more widely. Costs and benefits need to be assessed if only to make the business case. Some of the benefits will be quantifiable in money terms – time saving, paper reduction, speed of operation, for example. Others will be qualitative and difficult to evaluate. They might include quality of work, improved working environment, better documents, etc. The most important consideration may be the willingness of staff to change their methods of working, so it is vital to keep the users involved and informed in the process of selection, and to provide adequate training. In larger firms user groups can be established and help desks will provide on-line assistance. For smaller firms there may be an advantage in joining the software vendor's user group and sharing experience with other users. These vendors should also have help desks and keep records of known problems so that their software can be improved with each release.

2.7.2 IT System Management

Specialised maintenance and management of computer systems is a subject for computer scientists, but few small companies in construction employ people trained in this field. Even individual users of PCs need to know something about the maintenance of their systems and how to secure their data. Companies with more than 5 or 10 PCs will probably have a network and this needs some central maintenance and back-up. Valuable data should be duplicated on tape or CD-ROM every day or, at least every week, and copies kept in fire-proof safes and outside the office. One architectural practice had its CAD system stolen in the middle of a critical project and, although it was insured, inability to access the data for some time led to the failure of the practice. A common solution is for companies with similar types of computer system to have a mutual agreement for each to make their system available to the other if problems occur with one of them.

System security is a growing problem with an explosion in the numbers of viruses sent with e-mails. Good anti-virus software, such as that from Norton or McAfee, should always be used and updated regularly via the Internet. Some companies only connect one computer to the outside world and have a 'firewall' between it and the main network. A list of potential threats to computer-based information systems would include:

- Hardware failure
- Software failure
- Personnel actions
- Terminal access penetration
- Theft of data and equipment
- Fire and electrical problems
- User errors
- Program changes
- Telecommunications problems.

Of greater relevance to business management is the strategic effect of IT (Betts, 1999). This has been discussed in the context of Business Process Reengineering and Knowledge management, and IT is currently one of the most potent agents for change in an industry that can be rather resistant to change. E-business is growing in spite of the failure of dotcoms that were over funded too quickly in the 1990s. It is introducing many new techniques for trading in addition to the Project Webs already described. Some of these may not be welcome. Reverse auctions, in which bidders are invited to compete over the Internet during a short period after pre-qualification, could lead to even more exploitation of lowest price tendering. An example from an American state which offered a year's supply contract for school furniture by this method, found that, for the sake of being the first to win such a contract, firms were prepared to bid well below the level at which they could make any profit. At a time when the Egan philosophy is moving towards bidding on value, by taking into account quality as well as price, these e-business techniques could have an adverse effect.

2.8 ON-LINE RESOURCES

Business Management

http://www.sap.com/
SAP is one of the world leading business IT solution providers.

http://www.oracle.com/
Oracle provides IT solutions for business intelligence, collaboration tools, e-business, supply chain management, and so on.

http://www.unisys.com
UNISYS provides comprehensive IT solutions for businesses.

Electronic Document Management

http://www.dmcplc.co.uk/
ScanFile is an modular EDM system. It allows tailor made solutions to be built for specific application requirements. It supports barcode OCR indexing, Document Routing, Audit Trails, Forms Recognition complimented by WebServer and Remote Station capabilities.

http://www.acrosoft.com/
AcroSoft provides AS Documents an electronic document management system.

http://www.hummingbird.com/
Hummingbird DM delivers a comprehensive, enterprise-ready platform to harness and manage document-based knowledge assets.

http://www.lotus.com
Domino.Doc family, from Lotus, improves an organisation's efficiency through enhanced collaboration and information management. Domino.Doc delivers scalability, flexibility and low cost of ownership required to support both enterprise-wide document and records management.

http://www.pearldoc.com/
PearlDoc is a software services and development company, specialising in electronic information gathering, storage and retrieval software.

http://www.softco.com/
SoftCo Enterprise is an integrated document and process management suite.

http://www.powervisionsw.com
PowerView is an enterprise-class Integrated Document Management product suite that lets the user capture, organise, index, store, secure, access and transfer corporate documents.

http://docu-track.co.uk
DocuTrack offers a secure Document Management archive for documentations over an internal LAN or Intranet. It supports multiple file formats.

http://www.thirdverse.co.uk/
Paper trail is a document management program suitable for both individual and business users.

E-commerce Technology

http://www.buildonline.com
This company has developed an integrated suite of Internet based software solutions for the construction, infrastructure, utility and PFI industries. These include: ProjectsOnline for collaboration between project teams, TenderOnline for managing contract documents and SupplyChainOnline for managing the supply chain and tracking performance.

http://www.biwtech.com
The Building Industry Warehouse includes the usual news items on construction and a free directory of products, contractors, etc. It is primarily a project web

service but can also help in planning projects, and has plans to develop intelligent components and apply the Process protocol to construction.

http://www.uk.bidcom.com
Bidcom UK Ltd is part of Citadon company, who provides ProjectNet – an online software solution for collaboration on the design, construction and operation of capital projects and facilities.

http://www.4projects.com
4Projects focus on providing web-based database solutions enabling collaborative working. 4Project Extranet makes the very latest information available to project teams at any time of day or night and from any location.

http://www.aspectinternet.com
The Aspect Group specialises in the development, integration and support of Web-based applications and can deliver solutions spanning supply chain and retail management to asset management and training applications.

http://www.uk.easynet.net
Easynet is a leading pan-European commercial Internet Service Provider and telecommunications company with operations in eight European countries. In the UK Easynet has a national broadband network providing all levels of Internet access for clients.

E-commerce Services

http://www.cite.org.uk
CITE is a collaborative electronic information exchange initiative for the UK construction industry. Exchange formats are provided alongside practical support.

http://www.thenbs.com
The National Building Specification is a library of clauses for selection and editing to produce project specifications. It helps the specifier to tackle the complexities of the many types of construction involved in a building project.

http://www.construction-index.com
Abacus Construction Index is a directory of recommended online resources for UK architects, engineers and other construction professionals.

http://www.eu-supply.com
EU-Supply is the leading Internet-enabled procurement services organisation in Europe, serving over 80 of the top 200 construction businesses across 9 countries

http://www.4specs.com/
This is a portal for construction products in the USA. It has access to over 11,000 manufacturers' websites for construction product specifications, CAD details and CSI-formatted specifications.

http://www.4specs.com/
This is a portal for construction products in the USA. It has access to over 11,000 manufacturers' websites for construction product specifications, CAD details and CSI-formatted specifications.

Knowledge Management

http://www.autonomy.com
Autonomy technology provides a number of functions for organisations to manage their knowledge assets. Automatic categorisation enables high-level control, management and visibility of the knowledge base. Stored knowledge can be retrieved using a full range of retrieval options, from contextual, natural language queries, to legacy Boolean search. The automatic hyper-linking of related information ensures that users are fully aware of all the information assets available. Using Autonomy, content can be contextually summarised automatically and accurately, thereby enabling users to determine instantly whether a piece of information can help them. Autonomy's technology not only forms an understanding of content, but also of the people who read, create and process information. Implicit and explicit profiling means that content can be effectively personalised to the employees who need it to conduct their jobs efficiently.

http://www.kmworld.com/
This website provides on-line knowledge management publications and related events.

http://www.inference.com/
eGain provides knowledge management and e-commerce solutions. The eGain Knowledge Gateway supports questions/answers interaction in natural language. The eGain Knowledge Agent, using case-based reasoning, enables novice agents and trainees at helpdesk to cover a much broader domain than would otherwise be possible, and to successfully handle a significant percentage of inquiries that would otherwise require escalation. The eGain Knowledge Self-Service empowers customers to quickly resolve their problems at the instant they need help by walking them through targeted questions to pinpoint the best solution.

http://www.cogitoinc.com/
Cogito's products include a Knowledge Centre and suite of applications that shift the emphasis away from document-centric information management to knowledge-centric enterprise management. Assimilation applications allows information from many different sources to become part of a Knowledge Centre. Learning applications automatically learns how to reuse information in a Knowledge Centre. Interface applications assist authors with intelligent assistance.
View applications reuse knowledge in a Knowledge Centre to automatically generate documents required to support business processes. The Engine provides the platform for these and other next generation applications.

http://www.knowledgemanagement.uk.net/
This website is dedicated to research and practice of knowledge management (KM) in UK organisations as well as in academic and research institutions, with particular emphasis on the construction industry.

Government Initiative

http://www.ukonlineforbusiness.gov.uk
UK online for business is a DTI-led partnership between industry and government that helps all businesses make the most of their investment in information and communication technologies. It offers easy-to-use information and advice to help UK companies make the right decisions for their business.

SUMMARY

Computing is common to all industries and management techniques do not vary greatly between them. There are some processes specific to construction and much specialist software, particularly for construction management by contractors. Looking at the achievements of other industries in using strategic management to change the way in which they work, or to enter new business areas, could be useful to consultants and contractors alike. Much software is adaptable to suit the needs of specific industries. Standard systems should be the starting point when the business need for a system has been identified. Specially written software is still very expensive to produce and can take longer than expected to become fully deployed.

In the past businesses often acquired computer systems for the sake of learning about them or impressing their customers. With universal use and knowledge of computers this is not a valid argument, and each system should be fully evaluated before capital, and more importantly people time, is committed to it. The users should be consulted and prepared for changes in their working processes. Their comfort and training need to be addressed. It is amazing how architects and engineers crouched over drawing boards in the past and suffered the inevitable back problems but, as soon as they were given CAD workstations, any slight discomfort of seating position or screen glare was immediately corrected.

The manager's job is to know enough about the capabilities and weaknesses of IT systems and how they relate to the company's business, to be able to create a successful integration of these with the cooperation of staff, and to be able to see new possibilities and transform the business with the help of IT systems.

DISCUSSION QUESTIONS

1. In process change, what are the threats of substitution in construction?
2. Suggest some evolutionary and revolutionary changes for construction processes.
3. Give examples of tacit and explicit knowledge in construction.
4. What are the main benefits of e-business?

CHAPTER 3

CAD and Visualisation

LEARNING OBJECTIVES

1. Understand the principles and main functionality of Computer Aided Design (CAD) systems.
2. Appreciate how CAD helps to improve the quality and productivity of the building design process.
3. Understand the process of visualisation and its application to the construction industry.
4. Recognise a range of CAD and visualisation applications.

INTRODUCTION

Building design is a creative activity. The design team needs to create an imaginary building and to demonstrate it satisfies the client's needs. During this process, the designers must predict the building's appearance and performance when it is built. They also need to communicate the proposed design solution to the clients, and fellow team members. Traditionally, all these activities were done using design drawings and specification documents and by applying rules of thumb and simple algorithms. Since the early 1960s, there have been continuous efforts to develop computer systems to assist, not to replace, designers during the design stages.

The idea that the designer's ability could be enhanced with the aid of computing technology has been a major driving force for the development of computer aided design applications. People are motivated by the prospect of combining the creative and imaginative power of the designer with the analytical and computational power of the computer to make the architectural design process more informative and productive. As long ago as 1970s, Bazjanac (1975) identified three areas where computer applications were able to offer major benefits:

- Computer drafting systems could free designers from distracting and unproductive activities, allowing them to concentrate on the creative aspects of design.
- Computer analytical systems could support design decision making by enabling designers to test and evaluate design alternatives rapidly in the search for an optimum solution.
- Computer information management systems could offer designers instant access to accumulated knowledge in the building industry.

This chapter will concentrate on CAD and visualisation systems.

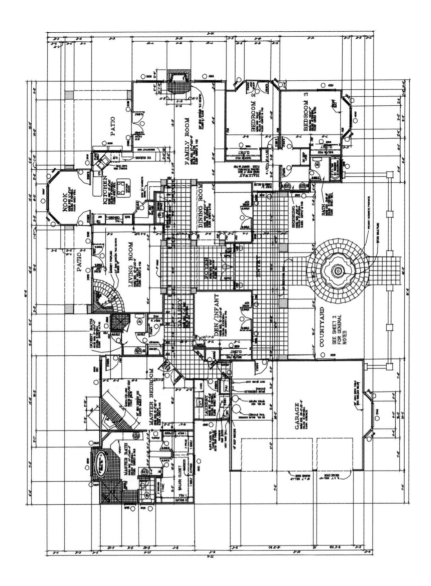

Figure 3.1 2D drawing (image courtesy of Autodesk)

3.1 COMPUTER AIDED DESIGN AND DRAFTING

Traditionally two dimensional (2D) drawings have been the main means of communication between members of the building team. Designers use a combination of 2D drawings and specification documents to communicate the design to engineers so that the engineers can make calculations. Similarly, the on-site building workers use drawings as instructions on where and how to lay the building foundation, how tall a wall should be built, how a wall and a roof are joined. For a typical building project, there are hundreds of drawings. The production of these drawing is a main task during the design stage. In 1970, architects were estimated to spend about 30% of their time on drawing.

The basic function of CAD packages is that they allow a user to build up drawings by manipulating lines, circles, rectangles and texts interactively on the screen. Some architecture-specific CAD systems even provide graphical libraries of commonly used building elements – doors, windows and so on (Richens, 1990). A typical 2D CAD drawing is shown in Figure 3.1.

Not surprisingly, the earliest CAD applications were those aimed at producing 2D drawings. A 2D drafting system can perform all the tasks that designers need to do using the traditional drawing board. These systems use simple graphic entities, including points, straight lines, curves, circles, polylines, rectangles, etc.

Using 2D drawing programs, commonly used objects in the design drawings, such as doors, windows and furniture, can be drawn and stored as a graphic library. Users can copy from the graphic library when a particular object is needed. *Autodesk's Architectural Desktop* provides a large number of these objects. Similarly, if a segment, such as a standard window on a building facade is repeated, there is no need to redraw it. The *copy* function enables the designer to reproduce the segment very easily. In addition to copying a single object, it is also possible to copy a *group* of objects. The advantage of this feature is apparent when drawing a multi-storey building. If some of the floor layouts are similar, after the first one is drawn, drawings for other floors can be produced quickly by copying.

The initial drawing alone does not show clear advantages for CAD, since experienced designers can draw on the drawing board equally fast. The real strength of CAD lies in its ability to allow 'editing'. Once a graphic is drawn, functions such as delete, move, copy, rotate, scale, mirror, etc., can be applied to any part of it. Other useful CAD functions are available such as repetition of an element at equal distances along a line, around a circle or on a grid or matrix; extending lines, partial erasures of lines and insertion of fillets. (Richens, 1990). These tasks cannot be easily carried out using paper-based media without restarting from scratch. The easy editing features of CAD systems enable designers to explore more alternatives of building layout during design (Kharrufa, 1988). Furthermore, since the drawing can be saved at any stage, the designers are able to keep various versions of the building layout for later study. Once the geometrical information on the building design is stored in a CAD package, different views of the building can easily be produced (Figure 3.2).

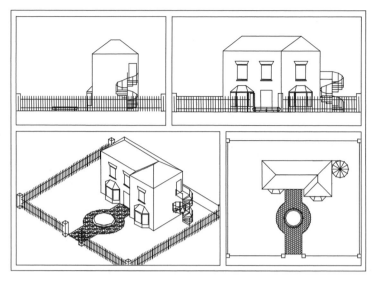

Figure 3.2 Multiple views of building produced by CAD

CAD started slowly with early systems designed for other industries, especially manufacturing. High cost and slow speed of hardware were limitating factors. In the mid 1980s when CAD systems started to run efficiently on PCs, their use began to grow rapidly. Costs have come down from about £50,000 per user in 1980 to the current (2003) starting cost of less than £5,000 for hardware and software.

For the last ten years there have been at least 50 different CAD systems addressing the needs of the construction industry. In recent years, some of the more competitively priced systems have tended to dominate the market with well packaged products; other companies have specialised in niche markets or providing a high level of support to their customers. AutoCAD, supplied by Autodesk, is the market leader in CAD applications. It is widely used by designers, engineers and consultants. Microstation from Bentley is popular with service engineers with its advanced multiple views of 3 dimensional models. ArchiCAD is well liked by some architects for its better visualisation functions.

Although there are some differences in the functionality of different CAD systems, they share some fundamental features. These are discussed in the following using AutoCAD as an example.

3.2 CAD FUNDAMENTALS

3.2.1 The Coordinate Systems

The CAD drawing area is a three dimensional electronic drawing space. It allows both 2D and 3D drawings to be drawn. The foundation for CAD is the coordinate system that allows a point to be specified in the drawing space accurately and consistently. AutoCAD uses the Cartesian coordinate system that has three axes: X, Y, and Z. The system coordinate origin is point (0,0,0). Every other point has three coordinate values of X, Y, and Z, that indicate the point's distance (in units) and its direction (+ or −) along the three axes relative to the origin (0,0,0), for example point (5,2,3) in Figure 3.3.

When a coordinate is relative to the origin, it is called an *absolute coordinate*. In practice, it is often difficult to work with absolute coordinates when drawing a complex 2D building layout or a 3D building model. In these cases, it is useful to relate a new point to a known point rather than the origin. For example, if you already know the coordinates of one corner of a window and the width and height of the window, it is very easy to work out the coordinates for all the other corners. *Relative coordinates* allow you to specify the X, Y and Z distance from a previous point. In AutoCAD, relative coordinates are entered using the @ symbol preceding the coordinate values. For example, if the first point is (5,2,3), a second point with relative coordinates of (@2,0,0) will specify the new point with absolute coordinates (7,2,3) as shown in Figure 3.3.

Polar coordinate is another coordinate system AutoCAD supports. It uses a distance and an angle to locate a point. To enter a polar coordinate, the user indicates a point's distance from the origin or from the previous point and its angle along the XY plane of the current coordinate system.

When a new drawing begins, AutoCAD is automatically in the world coordinate system (WCS); the X axis is horizontal, the Y axis is vertical, and the Z axis is perpendicular to the XY plane.

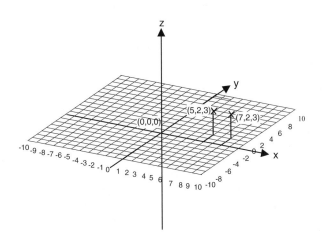

Figure 3.3 Coordinate system

3.2.2 Drawing Environment Settings

The characteristics of a drawing are defined by its environment settings that include drawing units and limits, snap and grid settings, and layer, line type, lettering standards, and so on. AutoCAD has default values for these variables when a new drawing is created. The user can modify settings to suit his/her preferences or the needs of a particular project.

AutoCAD coordinate units can be defined by different values according to particular requirements. For example, in one drawing, a unit might equal one millimetre of the real-world object. In another drawing, a unit might equal an inch. The user can set the unit type and number of decimal places for object lengths and angles. Drawing unit settings control how AutoCAD interprets coordinate and angle entries and how it displays coordinates and units in the drawing and in the dialog box.

3.2.3 Drawing Template

A template is a drawing file with predefined settings for new drawings. These settings include the size of the drawing area, unit type, border style and title block, and other parameters that can be predefined. It is useful if the user needs to produce a series of drawings of the same style, or if a company wants to have a standard style for all its drawings.

3.3 DRAWING WITH CAD

3.3.1 Creating Objects

AutoCAD allows the user to produce drawings by creating a range of objects, from simple lines and circles to spline curves, ellipses, and hatched areas. Figure 3.4 shows some of the simple 2D objects AutoCAD can draw. To draw an object, the user usually needs to specify a starting position and a series of coordinates. Of course, different types of object need different coordinate input data. To draw a line the user needs to give a starting point and an ending point. To draw a circle, the user needs to give a centre point and a radius. In many cases, there are different ways of drawing the same object. For example, to draw a circle, apart from specifying a centre point and a radius, the user can also draw the same circle with a diameter and two points. Other methods of drawing a circle include specifying a tangent to three existing objects or creating a tangent to two objects and specifying a radius. Skilled CAD users are usually able to choose the most appropriate drawing method for a given situation.

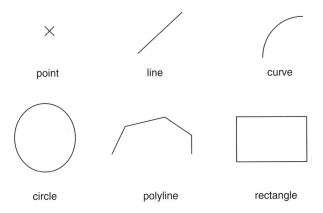

Figure 3.4 Basic 2D objects

3.3.2 Editing

AutoCAD offers a range of editing functions, including:

- *Copying*: This is a method of creating new objects by cloning existing ones. To copy objects within a drawing, the user needs to select the objects to copy either before or after issuing the *copy* command. Then specify a start point and an endpoint for the copy. These points are called the base point and the second point of displacement, respectively, and can be anywhere within the drawing. The *Copying* command can be used to make single copy or multiple copies of any existing objects.
- *Offsetting*: Offsetting creates a new object that is similar to a selected object but at a specified distance. This command can be used to offset lines, arcs, circles, 2D polylines, ellipses, elliptical arcs, etc. Offsetting circles creates larger or smaller circles depending on the offset side. Offsetting outside the perimeter creates a larger circle. Offsetting inside creates a smaller one.
- *Mirroring*: This command creates a mirror image of an existing object in the drawing. The new object is created using a chosen line as the mirror. The user can decide whether to delete or retain the original objects. Mirroring works in any plane parallel to the XY plane of the current UCS.
- *Arraying*: This command copies an object or a selection set in polar or rectangular arrays (patterns). For polar arrays, the user specifies the number of copies of the object and whether the copies are rotated. For rectangular arrays, the user specifies the number of rows and columns and the distance between them.
- *Moving*: This command moves the position of an object in the drawing.
- *Rotating*: This command changes the direction of an object.

- *Erasing*: The *Erase* command deletes one, or a selection of objects, from a drawing.

3.3.3 Annotations

Annotation is important to architectural drawings. Designers use textual explanations to complement graphic drawings. AutoCAD allows texts to be added to any drawing in various ways. For short, simple entries, the user can use line text. For longer entries with internal formatting, the user can use multiline text. The user can also format the font styles and sizes of the added texts.

3.3.4 Blocks and External References

A block is a combined object that contains one or more primitive AutoCAD objects. With blocks the user can organise and manipulate many objects as one component. The user can also associate items of non-graphical information with the blocks, for example, adding part numbers and prices by attaching attributes. Specification sheets or bills of materials can be created using this information.

An external reference (xref) links another drawing to the current drawing. When inserting a drawing as a block, the block definition and all of the associated geometry are stored in the current drawing database. The block is not updated if the original drawing changes. If a drawing is inserted as an external reference, it is updated when the original drawing changes. This function is very useful to create a number of related drawings.

3.3.5 Grid and Snap

Snap and grid are drawing aids that help the user create and align objects. The user can adjust the snap and grid spacing to intervals useful for specific drawing tasks. The grid is a visual guide displaying points at user-specified intervals, like customisable grid paper. Snap spacing restricts cursor movement to specific intervals. When the Snap mode is turned on, the cursor 'snaps' to spaced coordinates as if they were cursor magnets.

3.3.6 Layering

Layering is another powerful feature of a CAD system. Layers are like transparent overlays on which the user can organise and group different kinds of drawing information. The objects on the same layer have common properties including colour, linetypes, and lineweights.

A complex building plan drawing can be better organised by putting different objects on different layers. For example, all the structural elements may be on one layer, all interior furniture on another, dimension information and annotation on a third layer, and so on. Depending on the purpose, these layers can be overlaid on each other to produce a complete drawing or certain layers can be turned off for clarity. The effect of layering is shown in Figure 3.5.

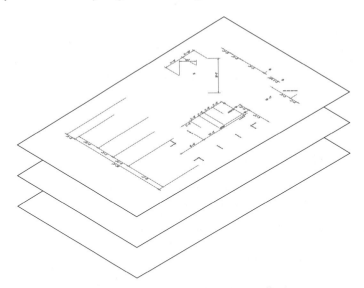

Figure 3.5 The use of layering in 2D drawing

There are standards for allocating drawing data to layers. A British Standard was first published in 1990 to suggest how layers could be allocated for building drawings. This recommended using the classification systems that were well known in building, such as SfB Table 1 or Common Arrangement, and using as few layers as possible when exchanging data. It was revised in 1998 to include the use of reference files and objects, and an International Standard, ISO 13567 was also produced. Each of these standards included a number of mandatory and optional fields, with the UK standard being a subset of the international one. An example of layer coding from BS 1192 Part 5 is shown in Table 3.1.

Table 3.1 An example to illustrate a layer name using all fields of the British standard

	Mandatory			Optional			
Field	Agent	Element	Presentation	Sector	Status	Scale	User defined
Name	A	244-	D	02BD	-	E	STAIRS
Description	Architect	Spiral stairs	Dimensions	Level 2, block B, zone D	Unused	1:50	Stairs

Layer conventions are quite easy to use and can be interpreted in different ways by each user. For coordinating project documents there needs to be agreement about a range of procedures – for example, how many layers will be used and which consultant will be responsible for which layers. This management requirement may be taken over by project Extranets, or Project Webs, in future. Where one consultant or an portal is responsible for holding project documents, it will need to establish guidelines for all participants to follow. These services could be very helpful in the integration of building project information.

3.3.7 File Format

The native AutoCAD file suffix is .dwg. Drawings can also be saved in .dxf format. DXF is a text file containing drawing information that can be read by other CAD systems or programs. AutoCAD provides many options for importing or exporting other file formats, including DXF, ACIS, WMF, BMP, and PostScript.

3.4 BENEFITS AND LIMITATIONS OF 2D CAD

Key benefits of 2D CAD include:

- 2D CAD tools increase productivity, especially at the information production stage. This benefit is more evident when design changes. Instead of reproducing all the drawings, CAD allows designers to make changes to the existing ones.
- 2D CAD tools help to improve the quality of design information. Using traditional paper drawings, keeping data accurate and consistent is a big challenge. In CAD drawings, all measurements are precise.
- 2D CAD tools help to increase the speed of information exchange between project team members. Previously, it would take several days to send drawings through the post. Now CAD drawing files can be attached to e-mails or sent via the Internet instantly.
- 2D CAD tools allow designers to reuse previous drawings or part of the drawings.
- 2D CAD tools make the drawing storage and archive task a lot easier. Paper drawings present storage problems as they deteriorate, are usually very large and cause documentation problems. A CAD file stored in a structured directory on a PC prevents these problems from occurring, which will also ease quality assurance issues.

However, there are also some limitations in the existing 2D CAD packages:

- Most CAD packages are only suitable for drawing at the information production stage when all major design decisions have been made, not during the early design stages.
- Current CAD drawing packages usually involve mainly geometrical aspects of the design. A building is represented by points, lines and surfaces, not by walls, windows and rooms. Although some of the latest

CAD systems adopted a model-based approach, there is still a lack of integration with other cost and performance analysis software packages.

- The user interface of many CAD programs is complex. Designers often find it difficult to operate them effectively. As a result, they can sometimes be distracted from the 'design' task that they should really be concentrating on. In some cases, specially trained CAD operators have to be employed to manage and use the drafting system. Such a practice increases the difficulty of integrating CAD applications into the design process.

- While not undervaluing the benefits of computer aided drafting systems especially for information production purposes, their weaknesses are also apparent. CAD drafting systems are only drawing tools and they do not improve the actual design process.

Case study
Introducing CAD

Jerram Falkus Construction Limited is a family-owned business with a turnover of around £27 million, which has been operating in London and the Home Counties since 1884. Prior to the autumn of 1999 the company had produced all its drawings by hand. However, Jonathan Leach, general manager of the joinery division was aware that designers increasingly wanted CAD-produced drawings – especially on large projects such as the £800,000 joinery contract recently carried out by the firm at Disney Imagineering's new European Headquarters in London. There were two primary reasons for investing in CAD: being able to transmit drawings electronically to the designers, and improving the perception of the firm among clients and their advisors. Leach spent several years evaluating the benefits of investing in CAD. When a vacancy for a draughtsman came up in the division the firm decided the time was right to invest in a CAD system and to recruit a trained operator. The firm now has two CAD stations and all of the design staff have been fully trained in their use.

The company chose AutoCAD LT on the basis of industry recommendation. Leach has no doubt that the new CAD system has been a key factor in the growth of the business. 'It's bringing substantially more work into the company,' he admits. 'I have seen an increasing volume of tender documents that ask you to state what CAD system you've got and what version. I can only deduce from that you won't be considered if you don't have CAD'. Investing in CAD has also brought benefits to the rest of the company. Jerram Falkus' IT Manager, Darren Smith, says: 'We are now able to receive drawings direct via e-mail from architects and engineers - no matter where they are based in the world - and open them in CAD. It has definitely given us an edge in terms of turning around information'.

Source: UK IT Construction Best Practice Programme

3.5 THREE DIMENSIONAL (3D) MODELLING

One of the limitations of 2D drawings is that they try to represent a three-dimensional building on a two dimensional plane. Inevitably some of the information cannot be represented explicitly. For instance, there is no way to show a wall is the same wall on a plan drawing and a facade drawing. To capture this kind of knowledge, a three-dimensional geometry modelling method is required. The progression from 2D to 3D CAD requires the use of more complex software. The graphics screen must now represent all three axes. In addition to creating and storing 3D objects, the CAD system needs the ability to generate different views of the objects. Depending on the way in which 3D objects are represented, 3D geometric models can be classified into *wireframe*, *surface*, and *solid* models.

3.5.1 Wireframe Model

Wireframe models are the earliest and simplest type of 3D model. In a similar way to 2D drawings, a 3D model represents a building, or any other object, using straight and curved lines. These lines show the edges of the model. There is no concept of *surface* in the wireframe model. Although wireframe models do not look like solid objects, they do contain an accurate geometric description of the object being modelled. See Figure 3.6.

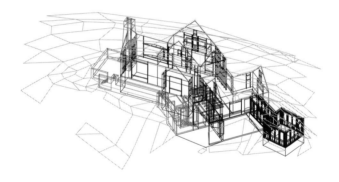

Figure 3.6 Wireframe model

Wireframe models have advantages of display speed, because the calculation involved is limited. Each and every line will be drawn, regardless of whether it is in front or in the background. However, a wireframe model does not have all the information about the object being modelled. Its display is often an ambiguous representation that leaves much of the interpretation to the user.

3.5.2 Surface Model

The second type of 3D modelling, the *surface* model, was first developed in the early 1960s. Surface models, unlike wireframes, provide both visual and mathematical descriptions of the surface shapes of the object. Surface models utilise the edge-vertex definitions of the wire-frame model, together with an additional list, the face list, in which the definitions of the edges used to define the surface patch will be stored in a specific order. Figure 3.7 shows a surface model.

An advantage of surface models is that they are easy to construct by creating plane surfaces, as well as by sweeping, revolving, or extruding entities. In addition, the designer can use patches to create a transition between adjacent surface edges. Surface models are also useful for finding the intersection of surfaces in space, volume calculation, mass estimation, and model creation for shaded renderings.

The primary limitation of a surface model is the lack of mass concept. As a result, a building element such as a wall cannot be represented by a block with a particular thickness.

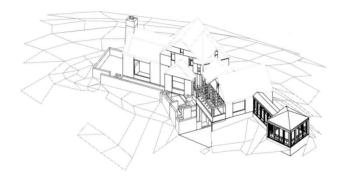

Figure 3.7 Surface model

3.5.3 Solid Model

A *solid* model, such as that shown in Figure 3.8, represents a building using both mass and its boundary surfaces. It is an unambiguous and complete description of the object being represented.

The construction procedure for solid modelling is different from that for wireframe and surface modelling. Instead of generating specific lines, curves, and surfaces that define the object, solid models use predefined solid primitives, such as blocks, cylinders, cones, wedges, spheres, and so on. The CAD user can define a particular user-primitive by specifying the desired shape, and then entering parameters such as size, position, and orientation.

Complex shapes can be generated by combining primitives using the following Boolean operations:

- union (the sum of two primitives)
- intersection (the common mass shared by two primitives)
- subtraction (subtracts a primitive from another).

Since solids contain more information about the closure and connectivity of shapes than wireframes and surface models, they have become the most important type of model for design, analysis, and manufacture of products. Solid models offer a number of advantages over surface models, including the ability to calculate mass properties such as weight and centre of gravity, which guarantees the structural integrity and accurate physical production of the model.

Figure 3.8 Solid model

3.6 VISUALISATION AND ANIMATION

With the advancement of more powerful computer workstations, users are now demanding methods to view these 3D models as shaded, photo-realistic images. Visualisation is the production of such images using computer systems. It usually involves two separate phases: that of *modelling* and *rendering*. In the modelling phase a designer creates a 3D model of an object using a variety of 3D modelling techniques discussed earlier. The result is a mathematical description of an object, typically described as a number of Bezier or B-Spline 2D curves and 3D patch surfaces. The techniques by which 3D geometric data is turned into photo-realistic (digital) images is called rendering.

3.6.1 The Visualisation Process

A typical visualisation process was classically described as five steps by Greenberg (1985):

1. *Three dimensional model*: The entire geometry of the environment must be mathematically defined as well as the colour and texture of the surface.
2. *Perspective transformation*: Each vertex of the model is transformed to generate a true perspective picture on the image plane.
3. *Visible surface determination*: Surfaces remaining within the frustum of vision after the perspective transformation are sorted in depth so that only the elements closest to the observer are displayed.
4. *Light-reflection model*: This model predicts the colour and spatial distribution of the light reflected from each surface in the environment.
5. *Image display*: The image is rendered by selecting the appropriate red, green and blue intensities for each pixel in the visible scene.

Visualisation aims at generating images which are not only correct but *real*. To achieve this, a number of factors need to be considered – the building geometry, its surroundings, material and surface texture, lighting sources and so on. A number of simulation methods have been introduced to produce effects that increase the realism of produced images including, reflection and refraction, diffuse inter-reflection and spectral effects.

Figure 3.9 A computer generated picture (image courtesy of IDPartnership-Northern)

3.6.2 Rendering

Through the application of a series of advanced mathematical algorithms, an artificial image can be created that looks as though it was shot by a real camera. Key terms used in the rendering process are:

1. *Hidden surface elimination*: Remove hidden lines and hidden surfaces from the image,
2. *Shading*: Add colour and simulated lighting effects to the surfaces,
3. *Texture mapping*: Simulate complex surface patterns using scanned-in or simulated textural detail,
4. *Shadowing*: Simulate real-world shadows which would be cast by objects.

Unlike the modelling process, rendering is computationally intensive because of the large number of floating point operations that must be performed to simulate the lighting, texture and shadowing effects. Until recently, photo-realistic rendering was restricted to the privileged few who had access to super-computers and high performance super-mini computers; in contrast, the rapid advance in CPU technology is now allowing this technology to become accessible to users of ordinary PCs and workstations alike. It is now possible to generate photo-realistic images on desktop computer systems.

The most popular techniques for calculating realistic images are ray tracing, radiosity and scanline.

- 'Ray tracing uses the principle that light rays will follow the same path in reverse if reflected back along themselves. For every dot on the screen, a ray of light is projected into the scene. If it should meet a plane, it will be reflected off it at an angle equal to its incident angle, but reduced in intensity depending on the reflectivity of the surface. If the plane should be glass or water, a refracted ray will also be produced at another angle. The rays are followed until they pass out of the scene or into a light source or are reflected more than a given number of times.' (Reynolds, 1993). This method is very good at simulating specular reflections and transparency, since the rays that are traced through the scene can be easily bounced at mirrors and refracted by transparent objects. While the ray tracing technique can be used to produce stunning images, its speed is very slow. This is because each ray must be intersected with every object in the scene. When the scene contains a large number of objects, the calculation becomes prohibitive on ordinary computers.
- The main idea of the *radiosity* method is to store illumination values on the surfaces of the objects as the light is propagated, starting at the light sources. There are a number of variations of this technique. Deterministic radiosity algorithms were the earliest radiosity method. Their speed is slow for calculating global illumination for very complex scenes. Stochastic methods were invented, that simulate the photon propagation using a Monte Carlo type algorithm. Galerkin radiosity is another method which tries to improve the speed of rendering. The difference between ray tracing and radiosity in the simulation is the starting point: ray tracing follows all rays from the eye of the viewer back to the light sources, radiosity simulates the diffuse propagation of light starting at the light sources.
- Another rendering technique is *scanline*. Scanline techniques operate by mathematically cutting the 3D scene with a 2D plane; this plane is aligned with a scanline of the monitor and passes through the viewer's eye location. The technique can quickly determine which surfaces are seen by

the viewer. Since the hidden surface information is determined on a scanline basis rather than on a pixel-by-pixel basis, the generation of the shaded image proceeds at a much faster rate, usually 5-10 times faster. The popular AutoCAD 3D Studio application uses the scanline technique.

3.6.3 Animation

In addition to producing single static images, some visualisation applications can produce a series of images. Then, these images can be displayed at a certain speed to generate the effect of a moving picture. This process is known as *animation*. Using this technique, a designer can show the client a pre-construction *walk-through* of the building. With accurate geometry, sophisticated colour, texture, material and lighting effects, animation systems can produce scenes with high-realism, so that designers and clients may explore the building spaces as if they were in the real building.

Figure 3.10 A VR model (image courtesy of Salford VE Centre)

3.7 VIRTUAL REALITY

3.7.1 VR Concept

Virtual Reality (VR) is a user-interface technology that allows humans to visualise and interact with computer-generated environments through human sensory channels in real-time. VR technology allowed the user to experience and interact with modelled objects in a virtual environment, as one would interact with the surrounding physical world. The user is no longer a passive receiver of computer-generated images. He or she can control the way in which the virtual world is explored.

The main keywords used in this definition are computer generated environment, interaction, human sensory channels and real-time. The meanings of these are given below (Fernando, 1997):

- *Computer Generated Environment*: The computer generated 3D graphical environment is the major component of VR. This environment is referred to as a virtual environment or virtual world. The virtual environment can maintain scientific data in scientific applications, engineering CAD models in engineering applications, CAD models of buildings in construction applications or a fantasy world in the case of entertainment. The objects within this virtual environment can be programmed to depict certain behaviour. Such behaviour can vary from rotating fans and real-time clocks to a car with realistic driving behaviour.

- *Visualisation and Interaction*: The ability to visualise and directly interact with objects in the virtual world is a powerful feature of VR. The type of interaction that the user wants to carry out within the virtual world varies according to the application. Such interaction can vary from a simple walk through to performing complex engineering operations such as assembly or disassembly tasks.

- *Real-time*: As the user is changing his or her viewpoint or interacting with the virtual objects, the virtual world needs to be continuously updated to give a sense of continuous motion to the user. The lowest acceptable frame rate in VR is 12 frames per second. For smooth simulations at least 24 (the frame rate of movies) or better, 30 frames/sec need to be displayed.

- *Human Sensory Channels*: The sensorial channels used in VR applications are visual, tactile and auditory. Researchers are also talking about the use of smell and taste. Perhaps in the future we may want to use smell to assess environmental pollution issues. However, technology for sensing smell and taste is not yet available.

3.7.2 Different Types of VR

According to the user interaction methods, VR systems can be divided into three types: non-immersive, semi-immersive and immersive.

Non-immersive VR
It is also called desktop VR, which includes mouse-controlled navigation through a three-dimensional environment on a graphics monitor, stereo viewing from the monitor via stereo glasses, stereo projection systems, and others. Another non-immersive VR example is Apple's QuickTime VR. It uses photographs for the modelling of three-dimensional worlds and provides pseudo look-around and walk-through capabilities on a graphics monitor. The main advantage of desktop VR system is their low cost in comparison with other forms of VR systems. However, a desktop VR system provides almost no sense of immersion in a virtual environment.

Semi-immersive VR
The term semi or partial immersive VR is used to describe projection-based VR systems. Reality Centre and Immersive Workbenches can be considered as semi-immersive VR systems. Projection based systems, consisting of a large, and often curved screen, and several projectors, provide a greater sense of presence than desktop systems because of the wider field of view.

Fully immersive VR
A fully immersive VR system is what most people would think a VR system should look like. To achieve full immersion the user has to employ a head-coupled display which is either head mounted or arranged to move with the head. A sense of full immersion is achieved because the display provides a visual image wherever the user is looking. Consequently, a head coupled display provides a 360 degree field view. All fully immersive VR systems give a sense of presence in the virtual environment that cannot be equalled by other VR approaches. This is a direct consequence of having a field view where images can be presented wherever the user is looking. The ability to exclude visible features of a real environment can lead to the sense of immersion taking place very quickly.

3.7.3 VR Display Systems

VR uses a range of sophisticated interface devices to create a realistic visual environment.

Head-Mounted Display (HMD)
A HMD uses two miniature screens that are placed very closed to the user's eyes, one for each eye, and an optical system that channels the images from the screens to the eyes, thereby, presenting a stereo view of a virtual world. A motion tracker continuously measures the position and orientation of the user's head and allows the computer to adjust the scene representation to the current view. As a result, the viewer can look around and walk through the surrounding virtual environment.

HMDs, which typically also include earphones for the auditory channel, have been the primary VR visual device for the 1990s. Their biggest advantage is their portability. However, due to ergonomic limitations such as weight, fit, and isolation from the real environment, many people find it difficult to wear a HMD for a long period.

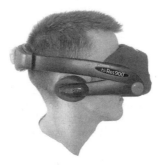 

Figure 3.11 Sample head-mounted display devices (images courtesy of Cybermind Interactive Nederland)

BOOM

The BOOM (Binocular Omni-Orientation Monitor) is a head-coupled stereoscopic display device. Screens and optical system are housed in a box that is attached to a multi-link arm. The user looks into the box through two holes, sees the virtual world, and can guide the box to any position within the operational volume of the device. Head tracking is accomplished via sensors in the links of the arm that holds the box.

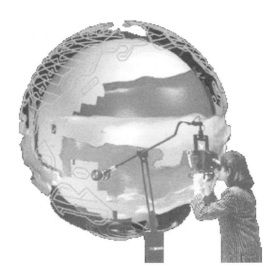

Figure 3.12 BOOM (image courtesy of Fakespace Systems Inc.)

Reality Centre

The term Reality Centre was first used by Silicon Graphics Inc. to refer to VR systems based on projection. The system consists of several, usually three, projectors and a curved screen. The screen provides a very wide view so that the viewer feels like as if he/she were inside the projected virtual environment. A Reality Centre allows a group of users to share the experience simultaneously.

Figure 3.13 Reality Centre (image courtesy of Fakespace Systems Inc.)

Workbench

The Workbench operates by projecting computer-generated, stereoscopic images off a mirror and then onto a table (i.e. workbench) surface that is viewed by a group of users around the table. Using stereoscopic shuttered glasses, users observe a 3D image displayed above the tabletop. By tracking the group leader's head and hand movements using magnetic sensors, the Workbench permits changing the view angle and interacting with the 3D scene. Other group members observe the scene as manipulated by the group leader, facilitating easy communication between observers about the scene and defining future actions by the group leader. Interaction is performed using speech recognition, a pinchglove for gesture recognition, and a simulated laser pointer.

Figure 3.14 Workbench (image courtesy of Fakespace Systems Inc.)

CAVE

The CAVE (Cave Automatic Virtual Environment) was first developed at the University of Illinois at Chicago and provides the illusion of immersion by projecting stereo images onto the walls and floor of a room-sized cube. Several persons wearing lightweight stereo glasses can enter and walk freely inside a CAVE. A head tracking system continuously adjusts the stereo projection to the current position of the leading viewer.

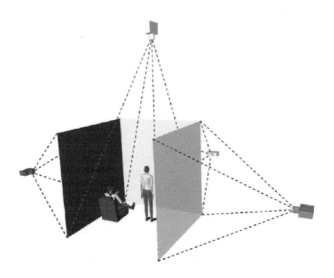

Figure 3.15 Illustration of a CAVE system (image courtesy of University of Michigan Virtual Reality Laboratory)

3.7.4 VR Input Devices

There is a variety of input devices such as data gloves, joysticks, and hand-held wands that allow the user to navigate through a virtual environment and to interact with virtual objects. Directional sound, tactile and force feedback devices, voice recognition and other technologies are being employed to enrich the immersive experience and to create more sensual interfaces.

Figure 3.16 Data glove and its interaction with a car VR model (images courtesy of University of Michigan Virtual Reality Laboratory)

3.7.5 Virtual Reality Modelling Language

The Virtual Reality Modelling Language (VRML) is the WWW manifestation of VR. It is a text-based file of definitions and geometry, specifying the composition of an image scene in 3D. This 3D scene can be viewed by a web browser equipped with a plug-in capable of understanding and rendering a VRML scene. The viewing of VRML models is usually done on a graphics monitor under mouse control and, therefore, not fully immersive. However, the syntax and data structure of VRML provide an excellent tool for the modelling of three-dimensional worlds. It is very cheap. The VRML browsers are usually free. VRML models can be easily integrated with HTML web pages (Figure 3.17). This makes it an increasing popular tool favoured by many users. The current version VRML 2.0 has become an international ISO/IEC standard under the name VRML97.

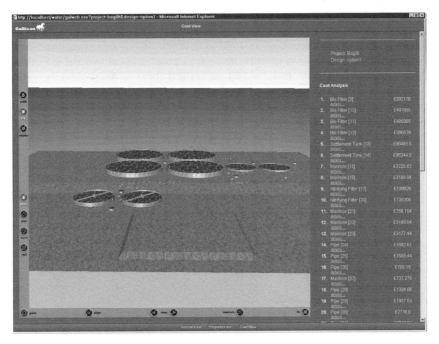

Figure 3.17 A VMRL model

3.7.6 Comparison between VR and Other Visualisation

In order to understand what VR is and is not, it is necessary to distinguish VR from related technologies. Fernando (1997) compared the three common types of computer visualisation systems, CAD, animation and VR:

- *Animation versus VR*: The main difference between animation and VR is in how the images are rendered. In animation, the images are pre-drawn, stored in a database and then played back in sequence. In VR, images are re-drawn in real time several times per second, in response to the user's changing position and actions within the virtual world. The main area of overlap between the two technologies is in the area of architectural walkthroughs and computer generated fly-pasts. However, in an animated scene, there is no possibility of instant interaction, or participation, between users and what they are seeing on screen. Whereas in VR, the user has the total control over what he wishes to see and where he wishes to go, within the virtual environment.
- *CAD versus VR*: VR surpasses CAD by being able to place users inside the model, allowing them to interact directly with the objects they are viewing rather than through a 2D computer interface. Many 3D solid model CAD systems allow the user to rotate an object but only within VR can the user walk around the object, stop, touch it, manipulate parts of it,

or even enter it. CAD software can be linked, via an interface, to VR software, in order to import data models into the virtual world. This data adds a powerful new way to understand and interact with CAD data. Furthermore, VR allows the user to improve the visual appearance of the products by applying lighting, texture mapping and surface properties. VR also allows the user to define the kinematic behaviour of objects to demonstrate the operation of a particular product.

3.7.7 VR Applications in Construction

VR is still fairly new technology. It offers great application opportunities for the construction industry. The *How to get started in Virtual Reality* guide, produced by Centre for Construct IT, outlined a list of tasks for which VR could be useful:

- Check and evaluate site access, particularly important for maintenance and emergency services.
- Plan and check access for the installation and use of large items of plant, particularly cranes, during the construction phase.
- Assessment of environments that are physically hazardous for human beings to enter.
- Provide an easily understood 3D view and link to design data held in conventional formats, such as web pages or images. This can be bi-directional.
- Visualisation of a proposed facility in its existing physical environment (a virtual photomontage).
- Monitoring of construction progress by integrating pictures (static or live) from site with the VR environment.
- Visual assessment of construction scheduling/sequencing. A lot of development work into what is known as 4D-modelling has taken place at CIFE, Stanford University, San Francisco and the Centre for Virtual Environments, University of Salford.
- Facilitate operations/maintenance and refurbishment/refits of existing facilities.
- Test 'what if' scenarios and present them as options in an easily understood manner.

There is no doubt about the potential benefits of VR for the construction industry. However, at present, high costs associated with both VR equipment and the process of creating VR models still preventing it being widely used in practice.

3.8 ON-LINE RESOURCES

CAD Software

http://www.graphisoft.com
ArchiCAD is favoured by many architects due to its good graphic and 3D modelling functions.

http://www.autodesk.com
AutoCAD is the market leader in CAD programs. In addition, Autodesk has a family of related CAD and visualisation products.

http://www.bentley.com/
Microstation is very widely used by architects and engineers. It combines CAD with 3D and visualisation.

http://www.builderscad.com/
BuildersCAD is a powerful CAD tool specifically for residential and light commercial buildings.

http://fastcad.com/n-home.html
Evolution Computing provides FastCAD and EasyCAD, both are extremely easy to use and cheap.

http://www.imsisoft.com/
TurboCAD is another easy to use and affordable 2D/3D CAD system. It costs less than a hundred dollars compared with several thousands dollars for AutoCAD and Microstation. Yet, it provides a range of 2D drafting, 3D modelling and visualisation functions.

http://www.intellicad.org/
TurboCAD is another easy to use and affordable 2D/3D CAD system. It costs less than a hundred dollars compared with several thousands dollars for AutoCAD and Microstation. Yet, it provides a range of 2D drafting, 3D modelling and visualisation functions.

http://www.acecad.co.uk/
AceCad Software Ltd is a leading supplier of software solutions to the structural steel industry. StruCad is a 3-D CAD system, which can create 3-D models of steel structures.

CAD Resources

http://www.cad-portal.com/
CAD-Portal is a website community that unites the best CAD resources on the Internet. The website focuses on CAD and GIS applications in manufacturing, engineering and AEC. It also hosts TechniCom's eWeekly newsletter, which focuses on the mechanical CAD/CAM/CAE/PDM marketplace. Each issue provides the latest industry news, events, and is often added to by industry analysts who provide reviews of the latest software and tools.

http://www.cadinfo.net/
This website has news, events and software guides on CAD systems. It also has free newsletters and magazines related to CAD.

http://www.cadstore.net/
An e-store gives you easy and instant access to specialised CAD and Engineering software. It also provides many electronic version of CAD books on a trial basis.

http://www.cadopolis.com/
This website has over 1000 AutoCAD related links, categorised for easy navigation.

http://www.architecturalcadd.com/
This website has a large collection of Internet resources and web links related to help and advice on CADD and design software for Architects.

Visualisation and Animation

http://www.solidworks.com/
Solidworks is a 3D modelling and visualisation tool.

http://www.abvent.com
Art•lantis Render is an easy and intuitive rendering program, which produces rendered images, animations and VR models.

http://www.spatial.com/
Spatial provides high-performance 3D software development technologies and services for 3D interoperability, modelling, and visualisation. SketchUp is simple, yet powerful software for creating, viewing, modifying and communicating 3D design concepts quickly and easily. AV-works is a ray-tracing and rendering engine fully integrated into ArchiCAD.

http://artifice.com/foyer.html
DesignWorkshop is a family of software power tools for creating 3D models, renderings, and walkthroughs, from initial sketches to polished presentations.

http://www.informatix.co.uk
Piranesi is a specialised '3D painting' tool, allowing you to start with a simple rendering of a 3D model, and quickly develop it into high quality images ready for client presentations.

Virtual Reality

http://www.vr.ucl.ac.uk
The VR Centre for the Built Environment at University College London.

http://www.nicve.salford.ac.uk/
The Centre for Virtual Environments at the University of Salford.

http://vicon.com/
VICON Motion Systems is a company that specialises in advanced VR applications in media, education and engineering.

http://www.fakespace.com
Fakespace Systems is a leading VR technology provider. It supplies a range of VR products, including WorkSpace (CAVE), WorkWall (Virtual Centre), WorkDesk (Workbench) and user interaction devices.

http://www.sgi.com/
SGI, Silicon Graphics, Inc., provides high-performance computing and visualisation technology. It offers VR equipment, such as Reality Centre, workbench and desktop VR systems.

VRML

http://www.w3.org/MarkUp/VRML/
VRML specification and links to VRML viewer software.

http://www.web3d.org/vrml/vrml.htm
The Web3D Repository is an impartial, comprehensive, community resource for the dissemination of information relating to Web3D, including VRML authoring tools and viewers.

SUMMARY

This chapter introduced CAD and visualisation systems. Many of these systems, especially 2D CAD, are widely used by architects, and more and more clients, contractors, and engineers have started using them as well. Several surveys in the UK have shown that systems of this type have a high penetration in the construction industry. They are often a basis for other types of construction IT applications. VR technology provides people with realistic experience in a virtual environment. It offers great potential for future use in the construction industry.

DISCUSSION QUESTIONS

1. What are the factors affecting growth of CAD systems in the construction industry?
2. How do the facilities of 2D drafting packages enhance the work of a designer? Consider editing, copying, layering, storage and transmission.
3. What are the benefits and limitations of each of the three types of 3D geometric models?
4. What are the relative merits of the different visualisation systems?
5. What are the main benefits of VR?

CHAPTER 4

Building Engineering Applications

LEARNING OBJECTIVES

1. Understand the nature of engineering design and construction.
2. Appreciate the benefits of simulating building performance.
3. Discuss the major types of building engineering applications.
4. Become aware of some environmental and structural analysis programs.

INTRODUCTION

Modern clients of the construction industry have increasingly demanding expectations. They want their buildings to look good visually; to be safe structurally; to provide comfortable living environments for their occupants; to consume less energy to operate; and so on. The ever more complex demands on the building design process have given rise to the need for a new approach to building engineering applications, based on computer software.

In this chapter, we outline some computer applications used by building services and structural engineers at the design stage of a project. We begin by examining the nature of engineering design and then consider the opportunities that computer software provides for simulating building performance.

4.1 ENGINEERING DESIGN

'The Engineer, inspired by the law of Economy and governed by mathematical calculation, puts us in accord with universal law. He achieves harmony.' *Vers une architecture. Le Corbusier.*

Writing in the 1920s, Le Corbusier (1927), the famous Swiss architect, contrasted the purity of engineering design with the over-elaborate forms then being used by architects. He looked to the design of ships and cars for the functional aesthetic that architects were only just starting to create. In any building project involving both architects and engineers, their different approaches to design have to be integrated. The architect is used to the gradual evolution of a design, even if it is driven by functionality, and exploring several possibilities. The engineer, although involved from as early a stage of the project as possible, usually will only want to carry out the detailed calculations required once, basing them on an agreed building layout. However computer software allows more alternatives to be explored and recalculation to be carried out quickly.

As engineering has become more complex, there has been greater specialisation into the main groups of civil, structural and building services, engineers. Many of the larger consultancies employ all these types and other specialists. Civil engineers play the leading role in large constructions such as:

roads, railways, bridges, harbours and airfields. They may be consultants or contractors and Isambard Kingdom Brunel was a pioneering example of a civil engineer who carried out both these roles in the 19th century. Specialisms that are most relevant to building are structural engineering, which is of even greater importance with recent threats to high-rise buildings, and building services engineering, which is further subdivided into electrical, heating and ventilating, mechanical and public health engineering. This chapter will focus on the use of IT systems in building services design. This is a growing field with many new developments, particularly in energy analysis and sustainability. Structural software has been essential for analysis and design for many years and is well covered in other textbooks.

There are ways in which buildings can be assessed for their impact on the environment that do not require computers. BREEAM, BRE's Environmental Assessment Method, is one of these and uses a series of checklists for particular types of building to rate their effects on: energy use, health, pollution, transport, land use, ecology, materials and water. The system covers offices, homes, industrial units and some other building types and these are rated as: pass, good or excellent. There are also methods using graphs to estimate environmental performance, but the interactions between heating, cooling, natural and artificial lighting, and with the systems that are installed to provide these, requires simulation models to examine their combined effects and find the best solutions.

4.2 BUILDING SIMULATION

The objective of Building Simulation is to predict, at the design stage, aspects of a building's performance such as its thermal, lighting and structural behaviour. It involves developing mathematical models to represent physical processes acting on the built environment. An example is the Building Research Establishment Domestic Energy Model that offers a simple simulation of energy consumption in domestic buildings. Other models simulate environmental performance based on hourly and monthly weather data and the effects of the service systems. Structural simulation allows loads, representing both typical and extreme cases, to be applied to a structure and its performance in bending, shear and deflection to be calculated. By applying these models at the design stage, designers can predict behaviour in terms of energy, lighting and structural safety.

The benefit of using building simulation is that potential building problems can be identified and corrected during design. It also enables designers to consider multiple performance aspects simultaneously to achieve a design solution close to the optimum. This requires rapid analysis of a range of options so that all the factors influencing a design can be balanced.

The built environment is extremely complex, so building simulation models are often very sophisticated. It is usually too time consuming to apply these models manually to a design project. During the last two decades, a range of computer software has been developed, based on simulation models. These computer applications are able to speed up dramatically the calculation process. These applications are typically classified as:

- Energy analysis systems
- Lighting analysis systems
- HVAC systems
- Structural analysis systems.

4.3 ENERGY ANALYSIS SYSTEMS

Before the 1970s, industrialised countries enjoyed relatively inexpensive energy supplies and energy saving was not a significant concern. The energy crisis in 1973 triggered a worldwide campaign to conserve energy. Buildings consume about 50% of all energy and this led to the construction industry being at the forefront of the campaign. Energy analysis is now an essential activity of building design and Part L of the building regulations requires calculations to be made.

Fundamental to building energy analysis are the thermal performance characteristics of external elements such as walls, floors, windows and roofs. Relevant sections of the CIBSE Guide define standard methods for calculating the thermal performance parameters, such as the insulation or U-values of external surfaces, their admittance (Y-value) or condensation risk. Computer systems based on these methods can perform calculations of the temperature gradient across a wall and predict the likelihood of condensation for user specified construction types, for example where the two lines cross in Figure 4.1.

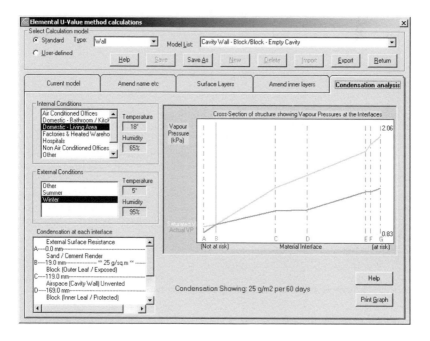

Figure 4.1 Condensation prediction within external wall (image courtesy of Owens Corning ACS)

Once the performance of an individual surface is calculated, the result can be compared with the Building Regulations requirements to see whether the performance of that particular building element is acceptable. However, steady state individual parameters can only give a rough indication of the performance of the building fabric. Quantitative evaluations for building heat losses, temperature variations and energy consumption require *dynamic* thermal analysis. Software of this type usually requires the designer to specify the building location, climatic data, thermal properties of the building fabric and various schedules for plant operation, occupancy, lighting etc.

Calculations can be performed to determine:

- hourly room temperatures
- monthly and annual heating/cooling loads
- monthly and annual fuel consumption of central plant
- monthly and annual total building fuel consumption and tariff analysis
- monthly and annual CO_2 production
- room surface temperature and heat fluxes.

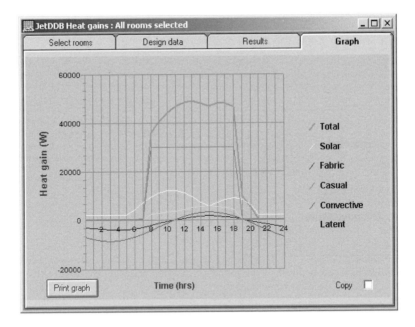

Figure 4.2 Sources of heat gains over 24 hours (reproduced with the kind permission of Hevacomp Ltd)

Case study
Avoiding Glare in Office

BT Castle Wharf office is a complex of linked office buildings of some 3,400 square meters on a city centre redevelopment site in Nottingham. To fully utilise daylight, the designers used substantial window areas on the building elevation and a centre atrium. There was a danger of glare problems in the office areas.

Integrated Environment Solution (IES) was employed to assess the potential problem and to give advice on lighting design. The first step of the analysis was to develop a 3D model of the building using ModelBuilder. This was achieved based on existing CAD drawings and technical documents. Then, SunCast and Radiance were used to simulate the sunshine and daylighting conditions in the building, especially in the areas surrounding the atrium. The simulation was able to calculate the luminance levels at different positions to identify excessive brightness that is the cause of glare. It was also able to determine the excessive contrast of light caused by inadequate distribution of daylight.

On the basis of this analysis, the IES advised the designers to use internal blind for certain areas in the atrium so that the building still benefits from good daylighting but glare was avoided.

Source: Integrated Environmental Solution Ltd

The graph in Figure 4.2 shows heat gains over 24 hours with the contributions made by solar gain, assumed to be on a sunny day in summer, gains through the building fabric, casual gains from lighting, cooking, etc., convective effects from the heating system and latent gains. This breakdown allows the sources of heat to be adjusted to avoid the need for special cooling systems after balancing with heat losses. Different times of the year and weather conditions can also be explored.

4.4 LIGHTING ANALYSIS SYSTEMS

Lighting analysis includes both daylighting and artificial lighting. Lighting conditions are an essential element of the built environment quality and have an impact on energy consumption. Good lighting design can maximise the use of daylighting to reduce energy costs whilst ensuring that areas within the deep core of the building are adequately illuminated artificially. Lighting also has an aesthetic function and proper design is important to convey the correct 'mood' within a building. Daylighting analysis is usually conducted on an individual room basis and takes into account the reflection from internal and external surfaces and the area of a standard CIE sky that is not obstructed.

For artificial lighting calculations, the designer needs to specify the room dimensional data, its surface reflectance, and the type and distribution of the lighting sources. The calculations can produce an average or point-by-point value of luminance to indicate the lighting level for that room. Alternatively, if the designer sets a specific illumination level, lighting analysis systems can calculate the number of luminaires needed to achieve this requirement.

Daylighting analysis can produce either a simple average daylight factor for a given room under the CIE Standard Overcast Sky condition or illuminance for points on a number of user-defined grids for given day and time.

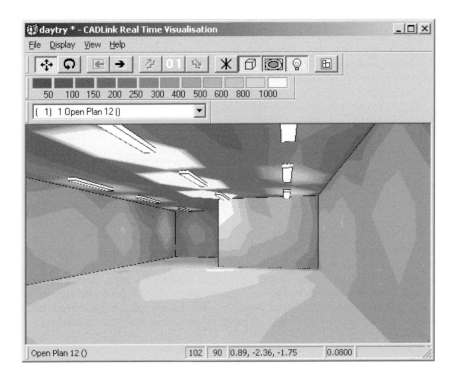

Figure 4.3 Lighting analysis (image courtesy of Cymap)

This example shows how the level of lighting can be represented visually. The artificial lights help to raise the lighting level at the backs of the rooms and it is possible to examine automatic lighting switching systems so that the artificial lighting is only on when the natural light falls below a level required for normal work of, say, 2% of external daylight.

4.5 HVAC DESIGN SYSTEMS

Heating, Ventilating and Air Conditioning design is the province of the building services consultants and contractors. The value of this work as a proportion of total building costs can be as high as 50% for highly serviced buildings such as hospitals. Environmental concerns have recently caused designers to try to minimise the amount of plant and to use natural stack ventilation instead of air conditioning where possible. Actual heat output from computers has been found to be less than the ratings given by manufacturers and ventilation loads in offices can be reduced as a result. Reduction in smoking over the last twenty years has also enabled designers to reduce the number of air changes required in buildings. The latest building regulations require testing of buildings for airtightness.

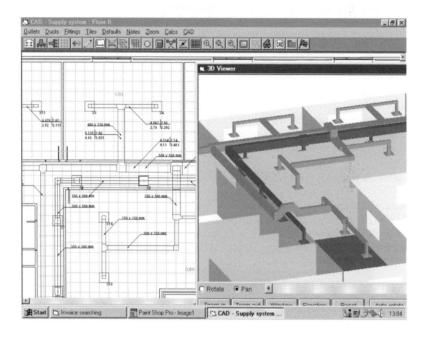

Figure 4.4 Mechanical design showing visualisation and drafting of ductwork (reproduced with the kind permission of Hevacomp Ltd)

Computer software is widely used for modelling the environmental behaviour of buildings at the early design stage. Critical areas such as atria are studied using fluid flow dynamics to ensure comfort at different levels. Housing now requires a Standard Assessment Procedure (SAP) rating which allocates points for energy efficiency. By the time the HVAC engineer starts design work, the fabric of the building has normally been fixed. Comprehensive energy modelling software such as *ESP* will model HVAC systems performance and that of the fabric and enable data on loads to be fed into the design process. Separate software packages are available for the design elements of particular service systems, such as those supplied by *Hevacomp* and *Cymap*. Both offer PC based software for the following types of application:

- *Mechanical services*: Plant design and sizing, Ductwork design and sizing
- *Piped services*: pipe network design and sizing, drainage and surface water
- *Electrical services*: cable sizing and load calculations, artificial lighting.

These packages, which are often sold as suites of linked software, may provide graphical aids to producing layout drawings. *Hevacad* is a package that selects appropriate items of ductwork and represents them diagrammatically. They can then be exported into *AutoCAD* to create full working drawings. Many of the systems use graphics to display charts of performance or even to represent services in 3D to visualise their appearance or check clashes of service with structure.

Another area of growing importance is control systems, which may be a part of more general Building Management Systems (BMS). Providing comfort conditions to suit individual areas of a building is now expected, rather than setting overall temperature and cooling levels. BMS provide central control and can also include security, fire and access control. Local outstations report data to a central station. Action may be taken automatically or by management staff. In future, software of increasing intelligence is likely to be developed using techniques such as neural networks, to interpret data and take action automatically. The long-term goal is referred to as the Intelligent Building.

4.6 STRUCTURAL ANALYSIS

Structural safety is a primary requirement of all construction projects. The objective of structural analysis is to ensure a building will support all the live and dead loads it is likely to encounter in its lifetime, and is safe. It also helps to reduce waste of materials by optimising structural design within safety margins.

Figure 4.5 Sydney Opera House – shells designed using structural analysis software by Ove Arup & Partners, engineers. Jørn Utzon, architect

Computer programs for structures can perform static and dynamic analysis of two- and three-dimensional structures on a variety of structural frames in any material. The input data for these programs include the building's geometrical outline, loading data and component and junction properties. The designer can usually enter these data using either graphical or tabular methods. Similarly the results of the analysis can also be presented in graphical formats.

4.7 ON-LINE RESOURCES

Organisations

http://www.bre.co.uk
The Building Research Establishment is the UK's leading centre of expertise on buildings, construction, energy, environment, fire and risk.

Case study
Modelling of Building Services

33 Old Broad Street is a £25 million 192,000 square feet office development in the City of London. HBG Construction was the main contractor for the project. The company is one of the top five contractor in the country. One of their assets is their IT capacity, which includes the ability to produce 3D models. Although they do not use 3D models for every project, it was considered appropriate to use 3D model on this project because of its complexity and high value.

Fulcrum Integration Consultancy was chosen to carry out the draughting, integration and modelling. The HBG Construction's building services manager also played a pivotal role during the model building process. He ensured the various organisations providing information for the model were brought into the project and provided the information and common when required.

In producing the 3D model, the first to be created was a model of the structure, then elements of architectural drawings, such as beams, walls, risers, staircases and openings, were added. Finally the services information was entered into the model from the scheme design information provided by the services engineers. The services included seven different components, ductwork, pipework, insulation, sprinklers, public health, controls and electrical. Particular attention was given to 'pinch points' where space was tight or which had particularly complex services and structure arrangement.

Once the model was created, the 3D software could perform clash detection analysis. The model also came a communication tool between all the stakeholders. The ability to visualise the services provided the means to make informed design decisions which led to more efficient co-ordination between different services and between services and structure.

Source: IT Construction Best Practice

---

http://www.bsria.co.uk
The Building Services Research and Information Association is an independent body providing advice, testing and research on building services and building physical performances.

http://www.cibse.org/
CIBSE, The Chartered Institution of Building Services Engineers, is an international body, which represents and provides services to the building services profession.

http://www.ping.at/cie/home.html
CIE, International Commission on Illumination, is an organisation devoted to international cooperation and exchange of information among its member countries on all matters relating to the science and art of lighting.

http://www.hvca.org.uk
The Heating and Ventilating Contractors Association is a membership-based body for companies who are involved in heating and ventilating systems.

http://www.ice.org.uk/
The Institution of Civil Engineers is an independent engineering institution. It was established in 1818, and today represents almost 80,000 professionally qualified civil engineers worldwide.

http://www.istructe.org.uk/
The Institution of Structural Engineers is concerned with the safety, efficiency and elegance of buildings and engineering structures.

http://www.ashrae.org/
ASHRAE, the American Society of Heating, Refrigerating and Air-Conditioning Engineers is an international organization of 50,000 persons with chapters throughout the world. The Society is organised for the purpose of advancing the arts and sciences of heating, ventilation, air conditioning and refrigeration for the public's benefit through research, standards writing, continuing education and publications.

http://www.nesltd.co.uk
National Energy Services is the trading subsidiary of the National Energy Foundation, an independent charity devoted to promoting energy conservation and awareness amongst consumers.

Best Practice Guide

http://www.bsbpp.org.uk
Building Services Best Practice encourages and facilitates business improvement in the building services sector, helping individuals and companies to achieve change, increase profitability and build competitive advantage.

http://www.actionenergy.org.uk
Action Energy, formerly Energy Efficiency Best Practice Programme, is a UK Government programme designed to provide free information to organisations to help them cut their energy bills.

http://www.buildingcentre.co.uk
The Building Centre provides a single source of information for the construction industry. It is a focal point covering all aspects of architecture and design, construction and planning, home improvement, DIY and self build.

Software Providers

http://www.hevacomp.com
Hevacomp provides a comprehensive package of building services design software for Mechanical and Electrical services and CAD. Mechanical design packages include Load calculations, Pipe & duct sizing and Mechanical CAD. Electrical design packages including conformance check with the requirements of IEE 16th Edition wiring regulations, Lighting systems design and electrical CAD.

http://www.cymap.com
Cymap is a member of the Graphisoft Group. It specialises in building services design software – both HVAC and electrical. Its flagship product, CADLink, is a comprehensive graphically-based product comprising a series of co-operating modules that can either be used together or individually. CADLink covers heating & cooling load calculations, energy consumption, pipe and ductwork analysis, psychometrics, public health, wiring and lighting design.

http://www.cads.co.uk
CADS provides a number of structural analyse programs, including SMART Engineer, A3D Max, SMART Modeller, SMART Portal, and CADS Analyse3D.

http://www.cscworld.com
TEDDS Calculation pad is design for the professional engineer to carry out structural analysis.

http://www.oasys-software.com
Oasys offers a suite of structural analysis packages: GSA – Frame and finite element analysis; steel member and reinforced concrete slab design; ADC – Analysis and design of reinforced concrete elements; AdSec – Analysis of general concrete sections for serviceability and ultimate limit states; and Compos – Analysis and design of simply supported composite concrete and steel beams.

http://www.reel.co.uk
Research Engineers International provides STAAD.*Pro* – a 3D structural analysis and design system, STAAD.*Pro* QSE – a quick structural analysis system, and ADL Pipe – a piping design system.

http://www.integer-software.co.uk
Integer Software provides a list of structural analysis software, including SuperSTRES, SuperSTEEL, SuperCONCRETE, SuperLOAD, and FEM.

http://www.masterseries.co.uk
MasterSeries provides a range of integrated structural analysis, design, drafting and detailing software for different types of structures.

http://www.acecad.co.uk
AceCad Software Ltd specialises in steel structural analysis systems.

http://www.flomerics.com

Flomerics provides simulation tools and services that enable an engineer to predict the behaviour of a proposed design prior to the build, called 'virtual-prototyping'.

http://www.edsl.net

Tas is a suite of software products, which simulate the dynamic thermal performance of buildings and their systems. The main module is Tas Building Designer, which performs dynamic building simulation with integrated natural and forced airflow. It has 3D graphics based geometry input that includes a CAD link. Tas Systems is a HVAC systems/controls simulator, which may be directly coupled with the building simulator. It performs automatic airflow and plant sizing and total energy demand. The third module, Tas Ambiens, is a robust and simple to use 2D CFD package which produces a cross section of micro climate variation in a space.

http://www.amtech-sys.co.uk

AMTECH is a leading software developer producing software specifically for the electrical industry.

SUMMARY

This chapter covered building engineering applications. It has concentrated on applications of engineering software in structures, environmental and services design. Engineers also use general applications described in other parts of this book and systems for civil engineering such as: foundations, hydraulics, roads, bridges, Geographical Information Systems and Finite Elements. All these systems should be used by suitably qualified engineers who take responsibility for applying their results. The output from computer programs should not be accepted without question since it is dependent on the appropriate use of the system and an understanding of how it works. Programs, such as the examples given, have good data checking facilities and use visualisation to confirm that the geometry looks correct. The other advantage of linking engineering analysis to CAD is that the sizing of services and structures can be transferred directly into drawings or building models, which reduces the risk of error. All construction professionals need to appreciate the importance of engineering software in ensuring the quality of building design.

Future goals for building engineering analysis are to develop complete integration between the different engineering disciplines, and with architecture and the construction process. This would involve creating 3D models from an early stage, sharing common data, such as geometry, and adding the additional data needed for engineering calculations.

DISCUSSION QUESTIONS

1. What are the benefits of Building Simulation and what advantages do computer software applications offer?
2. Why is the steady state thermal performance of a building inadequate to predict its heating and cooling requirements?
3. Why is daylighting an important aspect of Building Services design?
4. Discuss the advantages of linking building service applications.

CHAPTER 5

Computer Aided Cost Estimating

LEARNING OBJECTIVES

1. Understand the basic concepts of cost estimating and appreciate its importance for a construction company.
2. Recognise the benefits of computer aided estimating applications.
3. Discuss the features of cost estimating software.
4. Compare general-purpose spreadsheets with dedicated applications software for business and engineering.

INTRODUCTION

Controlling costs is one of the most important requirements of a construction project. To achieve this control, contractors and sub-contractors must first perform an accurate cost estimation to establish spending targets. Rigorous project accounting must then be employed to ensure that the actual spending will not exceed the target. Although it is possible to perform these tasks manually, computers can provide faster and more accurate answers.

We examine cost estimating, accounting and cost engineering as well as features of specialist accounting software and demonstrate how the use of spreadsheets has facilitated manipulation of data.

5.1 COST ESTIMATING PRINCIPLES

Cost estimates and cost control are essential tasks throughout the whole life cycle of a construction project. They are usually carried out by cost consultants or quantity surveyors. From the start of the project, a client needs to know the probable cost of the project, and thus set the budget for it. At this stage, because many factors are unknown, any estimates are usually indicative. At the design development stages, as the design is specified in more detail and more external conditions become known, the estimates of costs will become more accurate. The cost consultant can work with the designer to evaluate different design options from a cost perspective. At the tendering stage, estimators acting for potential bidders need to produce a reliable cost estimate low enough to win the contract and high enough to make a profit. At the project execution phase, estimating is needed to monitor project execution and audit project success (Akintoye, 2000).

Cost estimators may use a number of methods for preparing cost estimates for construction projects, ranging from an 'educated guess' to sophisticated analytical estimating techniques. Smith (1995) divided these methods broadly into two categories:

1. Estimating techniques using historical data
2. Unit rate estimating.

Estimating using historical data means predicting the likely costs of a new building using actual costs of similar projects from the past. The estimates are calculated per functional unit or per unit of floor area. It is a simple technique that is easy and quick to apply. The disadvantage is poor accuracy. This type of method is usually used at the early stage of a project when limited information is available.

Unit rate estimating is an analytical estimating technique that predicts costs by calculating the total amount of resources (labour, materials and equipment) required for a project and the costs of these resources. A list of these resources, known as a Bill of Quantities, can be developed using detailed working drawings and full specification of a design. This type of estimating method is more accurate and it is used at the late phase of design evaluation and tendering stages.

To perform a unit rate estimate requires gathering and analysing a large amount of data. Some of the tasks are time consuming and repetitive. Computers are ideal for providing support for these tasks. Therefore, the following discussion will focus on this type of application.

5.2 UNIT COST ESTIMATING

Traditionally, most construction projects were procured through competitive tendering where several contractors compete for the same project. Price is often the deciding factor in the award of a contract. It is crucial for the contractors to have some reliable methods of forecasting prices, and therefore likely costs, for future construction work (Smith, 1995). In recent years, new procurement methods, such as design and build and partnering, have emerged gradually replacing competitive tendering. However, accurate estimating of project costs is still important.

Unit cost estimating usually involves several steps:

- Quantity take-off
- Establishment of work methods and productivity rates
- Estimation of direct and indirect costs
- Compilation and analysis.

5.2.1 Quantity Take-off

Quantity take-off is the process used to produce the Bill of Quantities (BoQ) for a project. A BoQ is a detailed enumeration of all materials and components that go into a construction project, prepared according to industry standard methods of measurement, such as SMM7 (RICS, 1988). The task requires a quantity surveyor to complete the measurement from all the drawings and specification documents of the project. Computer programs can assist in the quantity take-off by helping estimators to measure, count, compute, and tabulate quantities, lengths, areas, volumes, and so forth, of objects found in the plans and specifications. One way would be to read drawings conventionally and type the data into a spreadsheet or custom take-off program. Some programs used in conjunction with a digitising tablet can read data directly from paper-based drawings.

5.2.2 Labour Costs

Labour cost is a major component of construction project cost. It is also one of the most difficult topics in preparing a detailed estimate. To estimate labour costs one needs to calculate the amount of labour required for the job and the unit cost of labour. The amount of labour is determined by the scale of the work and the speed at which the labour is able to complete it, or by labour productivity. There are a number of builders' price books published by, amongst others, SPON (2001) every year. These books provide guidance on 'standard' productivity for the most common construction tasks. Most estimators and estimating consultant firms usually have their own maintained databases. Computer software can be used to manage and provide access to production databases. It can also enable estimators to customise standard databases so that their own knowledge can be accumulated for future estimating tasks.

5.2.3 Materials, Equipment and Subcontractors

The quantities of materials come from the quantities take-off. However, a BoQ is usually measured by net requirement of materials without allowance for wastage. An extra percentage needs to be added to the estimated quantities. The prices of material can come either from industry standard price books or from quotations by specific suppliers.

Equipment cost refers to the hiring cost of special equipment needed for a project. The estimator needs to identify not only what equipment is required but also the length of time for which it is needed. Other costs include fuel, operators and transportation to and from the site.

In a construction project, the main contractor usually uses a range of specialist subcontractors for specific tasks, such as fitting of electrical systems, interior decoration, etc. In most projects a substantial part of the work is done through subcontracting, it is therefore important for the estimator to obtain quotations from subcontractors during the estimating process. Computer programs can help the estimator by subdividing parts of the work into packages that fit the craft structure and practices of potential suppliers, and track and analyse their quotes when they come in.

5.2.4 Indirect Costs

Contractors vary widely in how they define and handle the indirect and overhead costs in their estimates, and estimating software packages vary accordingly. Some costs, such as tax and national insurance burdens that go with the payroll, can be readily identified with the direct work. Even so some contractors prefer to total them and keep them as separate overhead items in their bid. For large equipment costs, estimators use a variety of techniques to spread the costs to each project. Other indirect costs are office expenses and consumables.

Good estimating packages offer flexibility to match a given contractor's preferences for dealing with overhead and indirect costs, rather than force them to change the way they estimate and bid for work.

5.2.5 Compilation, Analysis and Reporting

Estimating, leading to bid preparation, is usually a very hectic process especially in the final hours, as subcontractors and suppliers telephone and fax in their bids and prices for various parts of the work. The ability to keep updating changes is essential. Here a computer can really provide a pay off in its speed and accuracy. As the numbers come in, they must be rapidly analysed and checked, and decisions can be made on a well-informed basis.

Once the estimator has computed the direct and indirect costs, management must give consideration to matters such as the risks and contingencies associated with the work, the competition that will be faced in bidding, and how much profit to put on the bid. Computer programs allow managers to consider various options and answer what-if questions.

Cost estimating also needs to provide numerous analysis reports for the labour, material, equipment, and subcontractor costs, and summary reports to provide a better overall feeling for the bid based on the estimate.

5.3 COMPUTER AIDED ESTIMATING

Cost estimating involves many computation and numerical analysis activities. Computers are well suited for this type of task. Since the 1970s computers have been used to assist cost estimating. The earliest applications were in-house developed software for mainframe computers by some large contractors. Due to the high cost, they were not available to most estimators. In the 1980s PCs became widely used and the cost of computing declined rapidly. A new generation of purpose built cost estimating software emerged. Computer-based estimating programs are good at the data collection, computational, and clerical aspects of estimating. They archive and retrieve large volumes of resources, cost, and productivity information, perform calculations quickly and accurately, and present results in an organised, neat, and consistent manner. All these virtues are of tremendous help in the high-pressure environment in which most construction estimators often find themselves.

A survey in 1999 (Building Centre Trust, 1999) revealed that 44% of main contractors, 65% of sub-contractors and 68% of quantity surveyors, used computer programs for estimating tasks. The types of computer aided estimating applications fall into two broad categories:

1. Spreadsheet based bespoke programs
2. Off-the-shelf cost estimating packages.

5.4 SPREADSHEET APPLICATIONS

The first electronic spreadsheet application, VisiCalc, was developed by Dan Bricklin and Bob Frankston in 1979. It was one of the earliest practical applications on Personal Computers (PCs). It helped the wide acceptance of PCs by the business community. Succeeding spreadsheets, such as Microsoft Excel and Lotus 1-2-3, provide contractors and project managers with a powerful and convenient analytical and presentation tool.

Although spreadsheets can be used for presentation purposes, they are intended primarily for applications involving computations, particularly those that can be organised into rows and columns of text and numbers.

5.4.1 Basic Concepts

In this section, Microsoft Excel is used to discuss the basic concept of a spreadsheet application. A Microsoft Excel file is a workbook in which data can be entered and stored. A workbook contains one or many worksheets, and each worksheet contains a matrix of cells arranged in columns and rows (Figure 5.1).

The rows have numeric labels and the columns have alphabetic labels. A cell is identified by a label with a combination of column and row, for example F17 is the cell relating to column F and row 17 (Figure 5.1). In a cell label, or its address, the alphabetic column designation always precedes the row number.

The cell is the basic unit for storing data. Each cell can contain one of three types of information: a label, a value, or a formula (Capron, 2000). A label is a piece of text, which is usually used as a description of other data. A value is numerical data that can be used for performing calculations. A formula is a calculation using data from other cells. The real power of a spreadsheet lies in the use of formulas.

5.4.2 Constructing Formulas

Microsoft Excel and other spreadsheet software allow for interactive calculations involving multiple cells. They provide a wide range of built-in formulas for the users to develop their own applications. Formulas in Microsoft Excel follow a specific syntax, or order, that includes an equal sign (=) followed by the elements to be calculated (the operands), which are separated by calculation operators. Each

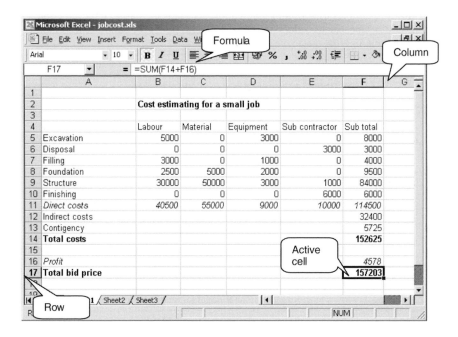

Figure 5.1 Spreadsheet user interface

operand can be a value that does not change (a constant value), a cell or range reference, a label, a name, or a worksheet function. For instance, in the example shown in Figure 5.1 the 'Total bid price' (cell F17) is calculated by 'Total costs' (F14) plus 'profit' (F15), thus cell F17 is calculated by the following formula:

=SUM(F14+F15)

The following format is used to specify a formula that operates on a range of cells along either a row or a column. The cell F9 contains a formula:

=SUM(B9:E9)

The advantage of using formulas rather than entering the value directly is that, if the original data such as 'Total costs' or 'Profit' are changed, the result of the formula, 'Total bid price', will be updated automatically.

Operators specify the type of calculation to be performed on the elements of a formula. Microsoft Excel supports all the common arithmetic operators including addition (+), subtraction (-), multiplication (*), division (/), percent (%), and exponentiation (^).

In the examples of formulas given above, SUM is the name of the function. It means adding the list of values together. In addition, Excel supports many other functions including the ones listed in Table 5.1.

Table 5.1 Commonly used spreadsheet functions

Function	Explanation
ABS	Return the absolute value of a number
AVERAGE	Return the average of the values in a list
CEILING	Round a number to the nearest integer or to the nearest multiple of significance
COUNT	Return the number of items in a list. This is the default function for non-numeric data
EVEN	Round a number up to the nearest even integer
FLOOR	Round a number down, toward zero
INT	Round a number down to the nearest integer
MAX	Return the largest value in a list
MIN	Return the smallest value in a list
MOD	Return the remainder from division
ROUND	Round a number to a specified number of digits
ROUNDDOWN	Round a number down, toward zero
ROUNDUP	Round a number up, away from zero
STDDEV	Return an estimate of the standard deviation of a population, where the list is the sample

5.4.3 What-if Analysis

The automatic calculation offers great advantages. For the estimating task it is common for some input data to change regularly. The fact that, by changing a parameter all results will be correctly updated, will save a lot of recalculation effort. It also offers the user the ability to evaluate different options by changing one, or a combination, of parameters. This is the principle of 'what-if' analysis. In the above example (Figure 5.1), the estimator might want to consider:

- What if the labour costs are increased to £6000 for Excavation by allocating more people to the task, but the Equipment cost can now be reduced to £1500 because the task can be completed more quickly?

- What if the indirect cost is increased from 80% of the Labour cost to 100%?
- What if the Finishing is done by in-house labour at a cost of £2500 for labour and £1500 for material?

With the spreadsheet, the estimator can try the above 'what-if' options and see the result of the 'Total bid price' instantly.

5.4.4 Graphic Outputs

Graphics have always been an effective communication tool in business. Cost estimators may be more familiar working with numbers. However, when communicating with managers and clients, graphics can help to get the point across more effectively. Furthermore, graphics are very good at showing trends of changes and ratios of different components within a total. Graphic data is called a chart in Excel. The commonly used types of chart include:

- *Area chart*: An area chart emphasises the magnitude of change over time. By displaying the sum of the plotted values, an area chart also shows the relationship of parts to a whole.
- *Column chart*: A column chart shows data changes over a period of time or illustrates comparisons among items. Categories are organised horizontally, values vertically, to emphasise variation over time.
- *Bar chart*: A bar chart illustrates comparisons among individual items. Categories are organised vertically, values horizontally, to focus on comparing values and to place less emphasis on time.
- *Line chart*: A line chart shows trends in data at equal intervals by linking the plotted data points of a series.
- *Pie chart*: A pie chart shows the proportional size of items that make up a data series to the sum of the items (Figure 5.2). It always shows only one data series and is useful when emphasising a significant element.

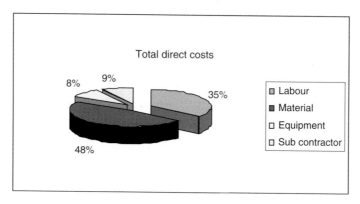

Figure 5.2 Excel pie chart

In addition, there are doughnut charts, stock charts, XY scatter charts, radar charts, surface charts, cone charts, etc. The simplest way of presenting the data should be used and 3D charts, such as the example below, may not be the easiest ones to read.

5.4.5 Customised Applications

Excel is very versatile and flexible. The user can customise its look and feel and create templates for common tasks. In fact, Excel provides a list of built-in templates that can be further customised by the user. Furthermore, it provides integral support for Visual Basic for Applications (VBA), so that bespoke applications can be easily developed by advanced users.

Even though Excel offers a rich list of functions, in some circumstances a user might need to extend the functionality through VBA programming, for example:

- Expanding the scope of Excel's functionality by adding new functions to perform specialised tasks. An example might be the collection of data, automating the collecting, data conversion, and generating reports.
- Simplifying a procedure. Some calculation procedures might be done in Excel through complex steps. In such a case, the user can develop a simplified function to improve the application efficiency.
- Developing a user-friendly interface. More intuitive interfaces can be developed using VBA for getting input from the user (Figure 5.3). The output can also be presented in formats that the user is more familiar with (Figure 5.4).
- Linking with external applications. Many business applications need data from different sources. For example, for cost estimating, the bill of quantities might come from a CAD drawing, and the unit cost data might come from a separate database. To achieve automatic data exchange between different programs requires data interface links to be developed. VBA provides a convenient solution for this.

5.4.6 Evaluation of Spreadsheet

Spreadsheets, such as MS Excel, have proved to be popular tools for many cost estimators. They have a number of advantages:

- *Inexpensive*: Excel comes as part of the Microsoft Office package. The whole package, including MS Word, PowerPoint, Microsoft Outlook, MS Access, as well as MS Excel, only costs about £300 in 2003. This compares with other cost estimating packages that usually cost several hundred to several thousand pounds.
- *Flexible*: Excel is very powerful and flexible. Its function can be easily customised and extended by the user.

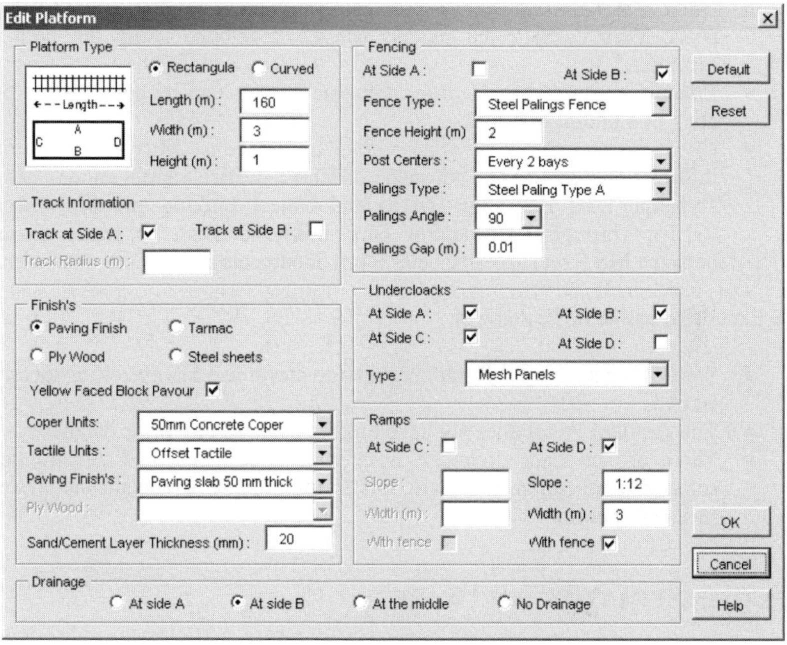

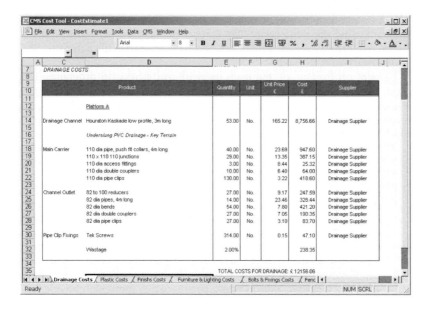

Figure 5.3 User interface of a bespoke spreadsheet program

Figure 5.4 A bespoke spreadsheet program

- *Good computational functions*: Excel is extremely good at computational tasks. Very complicated models can be developed using the built-in formulas.
- *Effective presentation facilities*: Graphic outputs can be generated at the click of a button in Excel.
- *Good connectivity with other Office applications*: Microsoft supports information exchange between different Office applications using Dynamic Data Exchange (DDE) and Object Linking and Embedding (OLE). Through DDE and/or OLE links, data can be easily shared between MS Excel spreadsheets, Word documents and Access databases.

MS Excel also has some limitations:

- Some programming skills are required to customise MS Excel for specific tasks.
- The standard Excel does not have links to external unit price databases.
- There are no data exchange links with other CAD and construction software application. However, this shortcoming is by no means limited to spreadsheets.

5.5 COST ESTIMATING SOFTWARE

A few large companies used computerised cost estimating programs on mainframe computers in the 1960s and 70s. However, due to high costs, their use was not widespread in the construction industry. With the rapid improvement of personal computers and the decline in cost in the last two decades, construction companies of all sizes can now afford to purchase computers needed to run estimating programs. During the same period, the number of 'off-the-shelf' cost estimating programs has grown considerably. Examples include MasterBill Estimator, Estimate, RIPAC, EVEREST, etc.

General contractors are the main users of cost estimating programs. Their purpose is to prepare estimates for project bidding. These programs allow an estimator to produce cost estimates quickly by linking BoQ with unit cost information from standard and user defined cost resource databases. Figure 5.5 shows a flowchart of the use of cost estimating programs.

There are four steps (shaded boxes in Figure 5.5) in using cost estimating programs:

- *Entering bills of quantities*: The first step is to enter a BoQ into the program. This can be achieved in a number of ways.
- *Linking to cost libraries*: This step links items of the bill to unit prices from standard or user libraries.
- *Analysing estimates*: At this stage, the estimators update the estimated costs by replacing some standard costs with accurate quotes from suppliers and subcontractors. They also carry out costs analysis by adjusting elements, such as waste, contingencies, profit margins, etc.

- *Producing reports*: After an estimate is finished, different types of report, e.g., bidding documents can be produced.

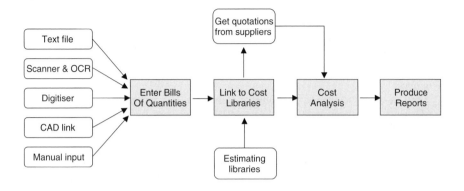

Figure 5.5 Flowchart of the use of a cost estimating program

These main functions are discussed in the following.

5.5.1 Electronic Bill of Quantities

The basic ingredient for a cost estimate is a bill of quantities for a project. Therefore, the first task in using cost estimating programs is to enter the BoQ into the system. This can be achieved in a number of ways:

1. Import from a text file
2. Input using optical scanner and Optical Character Recognition (OCR) software
3. Use a digitiser to take dimensions from drawings
4. Through a link with CAD software
5. Manual input.

Import from a Text File
Nowadays, most bills of quantities are prepared using computer programs of one sort or another. One way of exchanging BoQ is through a neutral format, such as the American Standard Code for Information Interchange (ASCII), Comma Separated Values (CSV) or fixed length data formats. The Construction Industries Trading Electronically (CITE) in the UK has defined a standard format for electronic bills of quantities. Many estimating programs, such as RIPAC and CONQUEST, can import BoQ files in these formats.

Scanner and OCR

In the construction industry most documents are still exchanged in paper form. Using a combination of scanner and OCR software, text and numbers on paper can be quickly converted to an electronic format. Since the 1980s, many estimators have used this technology in dealing with the input of bill of quantities. At present, the technology is still not 100% reliable. After scanning, editing is required to correct some errors. Nevertheless, it is much faster than retyping the whole document.

Digitiser

If a bill of quantities has to be produced from drawings, a digitiser is an ideal tool. CAD users have been using digitisers to take and enter measurement data for a long time. In recent years, some digitiser devices and associated programs, e.g., EasyGrid and EasyEarthwork, are specifically developed for the task of BoQ take-off.

CAD Link

Some programs can produce BoQ directly from CAD drawings. At present, most CAD drawings only contain geometric information. Even when textual annotations are used, these are not linked to the objects in the drawing. Furthermore, different designers use CAD differently. There are no standards for which symbols to use for particular types of object. Therefore, producing BoQs from drawings is not a fully automated process. Programs, like the CSSP CADLink, can navigate a drawing under the control of the user (Figure 5.6). The user will identify the objects on the drawing and the program will calculate the measurement.

Manual Input

All cost estimating programs allow BoQ data to be entered manually. In fact, this is still the most common practice amongst contractors. A BoQ is a list of bill items organised in a number of sections. For each item the information includes:

- Bill reference
- Work group code
- Item reference
- Item description
- Unit of measurement
- Quantities
- Cost rate
- Etc.

These data will be entered through a dialog box (Figure 5.7).

In many cases, the estimators do not need to type the whole of the bills from scratch. Instead, they can select existing items from standard libraries. The following is a list of libraries supported by many cost estimating programs:

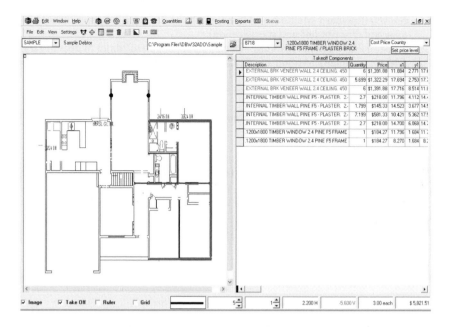

Figure 5.6 A graphic take-off program (image courtesy of Databuild)

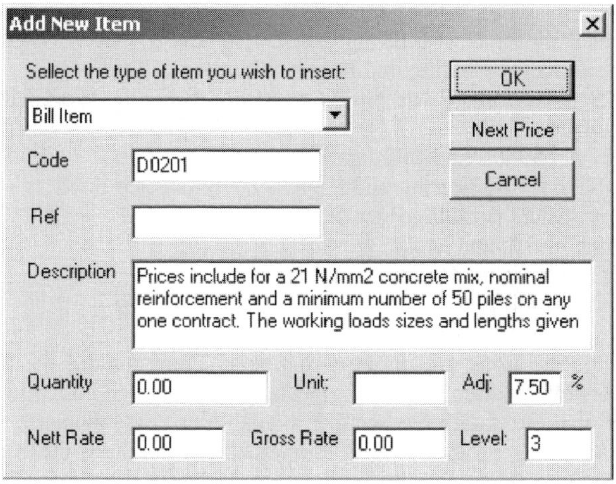

Figure 5.7 Dialog box for entering new bill items (image courtesy of Masterbill)

SMM7 (Standard Method of Measurement of building works, with 65,000 standard descriptions and specification)
SMM7 NBS (National Building Specification version of Major Works library)
SMM6 (Standard Method of Measurement of building works, Minor Work library with 5000 standard item)
M & E SERVICE (Mechanical and electrical services)
CESMM3 (Civil works library)
MMHW (4) (Method of Measurement for Highway Works)
NSR (National Schedule of Rates).

Each of these libraries provides a comprehensive list of standard items. Each of these items has its reference, code, description, and sometimes its unit price defined. All an estimator needs to do is to enter the quantities information specific to a project.

In addition to the use of standard libraries, estimators can also use data from previous projects.

5.5.2 Link BoQ to Cost Libraries

Standard component cost and/or unit cost of material, labour and equipment are essential for cost estimating. Traditionally, there are several well known and widely used industrywide information sources in the form of published price books, including:

Various Building Cost Information Service (BCIS) Price Books
Faber and Kell's Heating and Air Conditioning of Buildings
SPONS Estimating Costs Guide to Minor Building Works Refurbishment and Repairs
SPONS Architects and Builders Price Book
SPONS Civil Engineering and Highway Works Price Book
Wessex SMM7 Building Price Book
Hutchins' Small and Major Works
Griffiths' Complete Building Price Book
Laxton's Building Price Book: Major & Small Works.

These publications are updated regularly. They contain the latest industry standard costs information about all aspects of building and construction. In recent years, most of these data have become available as cost databases that can work with cost estimating programs. The estimators can purchase these databases and link their BoQ to the cost information to calculate the estimated cost for their projects.

The estimators can customise these standard resources by changing the cost rates so that they are more accurate for their specific context. They can also add new items.

5.5.3 Cost Analysis

Cost estimating is not a mechanical process. An accurate estimation requires knowledge and judgement of experienced estimators. Cost estimating programs cannot replace professional estimators. They provide support for routine calculations and instant updates in response to alternative resource costs, production rates, wastage factors and so on.

In a typical project, the main contractor uses many subcontractors and external suppliers for materials, labour and equipment. Their cost quotations need to be fed into the cost estimate. One of the estimating programs' functions is to coordinate and synthesise information from multiple sources.

After an estimate is initially produced, cost estimators can carry out cost analysis. Estimating programs allow the estimators to examine costs at different levels, project, section and individual item, by presenting an easy to read cost summary.

Case Study
An Integrated Design, Cost Estimating and Planning System

Corus Rail established a new business, Corus Rail Modular Systems (CRMS). The objective of CRMS is to design and build modular platforms for the railway industry with potential clients being Railtrack, and train operating companies.

The CRMS enable railway platforms of any shapes and sizes to be built using a small number of standard modular components. These components are manufactured in factories and assembled on site. The main advantages of CRMS over the traditional brick and mortar structures include shorter construction time, easier replacement and cheaper life cycle cost.

To improve the design process, Corus developed an integrated design, cost and project planning system. The system integrates AutoCAD, a customer built spreadsheets cost estimating system and MS Project. It is capable of generating designs for a given platform layout in CAD. The design data can then be exported electronically to the cost estimating system. A BoQ is generated automatically from the design data. Estimates of cost can be quickly worked out with unit cost information from a library database. The database contains suppliers' contact details as well as their products and prices. When the design is finished, the design data can similarly be exported to MS Project so that scheduling can be carried out. The dates from the finalised schedule are used to generate orders for the right products from the right suppliers at the right time.

5.5.4 Producing Reports

Once an estimate is completed, it can be either printed out in traditional tender document format (Figure 5.8) or exported to the CITE format so that tendering can be submitted electronically.

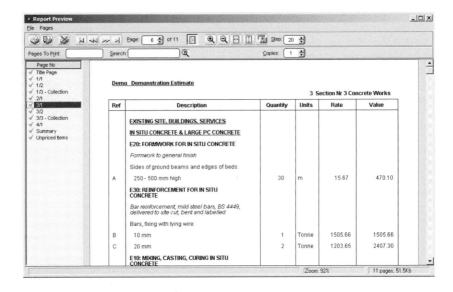

Figure 5.8 Cost estimate print out (image courtesy of Worldwide Software (UK) Ltd)

5.6 ON-LINE RESOURCES

Resource Libraries and Price Books

http://www.tiwessex.co.uk
Wessex price books, in two volumes – Major Works, Minor Works, contain over 2,000 pages packed with prices, 500,000 price elements, over 11,000 separately listed material supply prices, over 50,000 metric unit rates and over 5,000 approximate estimating prices for measured items.

http://www.pricebooks.co.uk
SPON Press publishes a series of price books, including Spon's Architects' & Builders' Price Book, Spon's Landscape & External Works, Spon's Civil Engineering & Highway Works, Spon's Mechanical & Electrical Services, Spon's Estimating Costs Guide to Electrical Works, Spon's Estimating Costs Guide to Plumbing and Heating, Spon's Estimating Costs Guide to Minor Works, Refurbishment, and Repairs, Spon's First Stage Estimating Handbook, Spon's

Manual for Educational Premises, Spon's House Improvements Price Book, Spon's Railways Construction Price Book and so on.

http://www.bcis.co.uk
The Building Cost Information Service (BCIS) Ltd & Building Maintenance Information (BMI) provides current, accurate information on UK building costs, tender prices and building maintenance. Subscribers can access cost information for a wide range of commercial, industrial, residential and public sector buildings. BCIS is for capital cost information while BMI covers maintenance management information and building maintenance, property occupancy and refurbishment costs.

http://www.barbour-index.co.uk
The Barbour Index Building Product Expert provides on-line access to building products and manufacturing information.

http://www.tionestop.com
Technical Indexes (ti) provides technical data, product/supplier information, and standards to construction industry worldwide.

Estimating Software

http://www.conquest.ltd.uk
Conquest is an estimating software package with fully integrated manual or digitiser take off and bill production. It supports a wide range of bill of quantities input methods, from scanning, ASCII files on disc, CITE files on disc or rapid manual input. It can be linked to libraries for SMM6, CESMM3, SMM7 & MMHW.

http://www.crestsoftware.co.uk
Valesco Estimating is just one of the Crest Construction Software products available to the Construction and Civil Engineering Industry. Valesco is designed to work from an imported BoQ or from a manually created BoQ using drawings/specification. Valesco is multi functional and by way of a common data file format allows a seamless transition from estimating to post contract status.

http://www.construct-it.co.uk
Construction Software Services are the developers of Construct-It – an estimating software package. It comes with a choice of Major Works (SMM7) / Small Works (SMM6) / Civils (CESSM 3) / Plumbing & Electrical / Suspended Ceilings for the UK

http://www.cssp.co.uk
The Construction Software Services Partnership offers several cost estimating packages, including RIPAC Estimating, ICEMATE, and CADLink.

http://www.masterbill.com
Masterbill is a complete QS system including BQ production, cost analysis, tender analysis, valuations, eTendering, and simultaneous remote working. QS CAD, coupled with Masterbill, can help to produce measurement from CAD drawings and scanned drawings.

http://www.jobmaster.co.uk
Jobmaster is a complete cost estimating, tendering and job management package.

http://www.estimate.co.uk
Esti-mate produces quotations and traditional Bills of quantities with prices created from Labour, Plant, Materials and Subcontractor resources. It works with industry standard price databases.

http://www.tiwessex.co.uk
The Wessex estimating tool allows for rapid and accurate pricing of work. The system has exclusive links to the Wessex Comprehensive pricing Libraries and Technical Indexes Product, British Standards and Health & Safety data.

http://www.csb.co.uk
Construction Industry Solutions (COINS) provides an integrated modular system comprising a number of components that can be assembled and implemented according to different needs. Cost estimating is one of the modules.

http://www.ecl.uk.com
QUICK EST provides the user with a fast method of producing an elemental estimate based on historical costs. The CATO Suite supports cost advice and management for all stages, from feasibility through to final account to costs in use. CADMeasure is a multipurpose measurement tool that allows the user to measure with accuracy from a CAD drawing.

http://www.causeway-tech.com
Causeway Estimating is a sophisticated estimating software package. It supports multiple standard industry databases for all sectors (SMM, CESMM, etc.). It is CITE (Construction Industry Trading Electronically) compliant and supports links to complementary services including Causeway's Cost and Works Management such as Siteman and the Causeway Directory, enabling electronic communication across the entire industry on every aspect of a project.

http://www.ramesys.com/construction
ESTEEM is a cost estimating software, which can run on a stand-alone computer or run over the Internet using a standard browser interface. Web enabled Esteem offers true remote site capability for users to produce accurate estimates. It supports scanned document imaging that allows the user to price against the original bill of quantities without the need for time-consuming Optical Character Reader (OCR) conversion. It uses industry standard pre-built libraries for budget and costing information such as SMM7, CESSM, roads, bridges and railways.

SUMMARY

Cost estimating is vital for construction companies in their bidding for contracts and ensuring each project is operating within budget and at a profit. IT applications enable the contractors and program manager to build up more accurate estimates of project cost by taking into account more factors in a detailed project breakdown and to monitor spending during a project.

DISCUSSION QUESTIONS

1. Why are spreadsheets the ideal tool for business and engineering?
2. Think of a problem in your daily life or in your workplace where spreadsheets can provide a solution.
3. What is the advantage for integrating cost estimating programs with cost databases?

CHAPTER 6

Planning, Scheduling and Site Management

LEARNING OBJECTIVES

1. Understand the principles of project planning and scheduling programs.
2. Recognise the importance of database for procurement systems.
3. Discuss the advantages of computer based surveying systems.
4. Appreciate the potential of operation simulation systems.

INTRODUCTION

It is a common misconception that computers are of little help on a building site because on-site operation is mainly physical work. In fact, construction work requires careful planning and skilful management of human and physical resources. Computer systems can assist on-site managers to plan ahead, evaluate different production options, adopt and execute the most efficient construction operation.

This chapter examines the use of computers for project planning and scheduling, procurement and surveying. We also consider the ways in which on-site operations can be simulated by computer to ensure that the various activities are properly synchronised.

6.1 CONCEPTS OF PROJECT PLANNING AND SCHEDULING

Although project planning and scheduling is not limited to the on-site work, it has particular importance during this phase because on-site operation involves the co-ordination of many activities. Project planning and scheduling involves defining all the activities of an operation and the duration of each activity, and the sequence according to which all activities are to be carried out. There are well established planning and scheduling methods and techniques.

6.1.1 Top Down Planning

Top down planning is a process of breaking a whole project into smaller components. In some cases, smaller components can be partitioned further into sub-components. Each component is a relatively self-contained work package and the interface between different components can be clearly defined.

6.1.2 Work Package Structure (WPS)

WPS, or Work Breakdown Structure (WBS), is a top down planning technique. It breaks a project into a hierarchical structure of working packages. For example, for a house project, WPS at the first level, and the second level of the first package, can be broken down into the following:

1. Site set up
   1.1 Erect fences
   1.2 Erect workers' cabins
   1.3 Install crane
   1.4 Connect electricity
   1.5 Connect water
   1.6 Other
2. Excavation
3. Structure
4. External walls
5. Partition walls
5. Services
6. Fit-out.

Of course, all the first level packages can be broken down into second level packages. The second level ones themselves can be decomposed further if necessary.

6.1.3 Programming and Scheduling

Once all activities are identified in the form of WPS, the next task is to put them in a logical order and determine the duration for each activity. This process is called programming and scheduling. Programming and scheduling are two different yet related concepts. Programming refers to the process of defining the sequence of tasks to be undertaken. It does not assign a time duration for each task. Typical programming techniques include activity lists, flow charts and networks, such as Critical Path Method (CPM) and Program Evaluation and Review Technique (PERT). Scheduling, built on the basis of programming, assigns the estimated time duration needed for each task. It involves measurement of quantities, productivity, and knowledge of the availability of resources (labour and equipment), and balances the need for resources at different stages of the project. Combinations of programs and schedules are typically presented as bar charts (Gantt-charts), line-of-balance charts, multi-activity charts, or timed networks. The three most widely used techniques, PERT, CPM and Gantt-charts, are explained briefly below.

PERT Charts
PERT charts contain task, duration, and dependency information. Figure 6.1 shows a simple PERT chart that starts with an initiation node from which the first task begins. In a PERT chart, a node represents an event or a point in time. A task is

represented by a line that states its name or other identifier, its duration, the number of people assigned to it and, in some cases, the initials of the personnel assigned. The other end of the task line is terminated by another node that identifies the start of another task, or the beginning of any slack time, that is waiting time between tasks. Multiple tasks can begin at the same time, in which case they all start from the same node then branch out from the starting point.

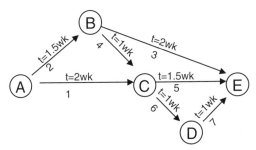

Figure 6.1 A simple PERT chart

All PERT charts must conform to the following rules:

- There is one starting node and one ending node. This starting node can only have outgoing lines. The ending node can only have incoming lines.
- The starting node represents the start of a project. The ending node represents its completion.
- A task (a line) leaving a node cannot start until all incoming tasks of that node are complete.
- There should be no cycles in a PERT chart. Otherwise, a system represented by the chart will be trapped in a never-ending loop.

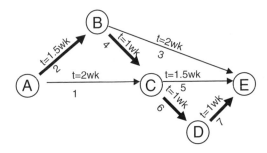

Figure 6.2 A CPM chart

CPM Charts

Critical Path Method (CPM) charts are similar to PERT charts and are sometimes known as PERT/CPM (Figure 6.2). A critical path consists of a set of dependent tasks which together take the longest time to complete. For example, in Figure 6.2 the critical path is indicated by the bold lines. Tasks, which fall on the critical path, need special attention because they are potential trouble spots.

Gantt Charts

Gantt charts, also known as bar charts, represent tasks using a collection of horizontal bars (Figure 6.3). The horizontal axis of the Gantt chart is a time scale, expressed either in absolute time or in relative time referenced to the beginning of a project. The time resolution depends on the project – the time unit typically is in weeks or months. Rows of bars in the chart show the beginning and ending dates of individual tasks in a project.

In Figure 6.3, some bars overlap. This means these tasks are carried out in parallel. Enhanced Gantt charts can also show task dependencies, as shown by the arrowed lines in the figure. The strength of the Gantt chart is its ability to display the status of each activity at a glance. This makes it the most widely used project planning tool in practice.

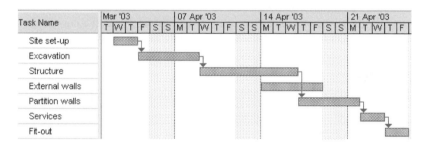

Figure 6.3 A Gantt chart

6.1.4 Resource Allocation

Another task during planning and scheduling is the allocation of resources (labour and equipment) to each task. A project planner needs to balance the task duration and resource allocation. For example, if a task needs to be completed more quickly, more resources need to be allocated. The planner also needs to consider the overall availability of resources at different stages of a project and allocate them appropriately to achieve the most efficient result.

6.2 BUILDING A PLAN USING PROJECT PLANNING PROGRAMS

There are many commercial computer programs for project planning and scheduling purposes. The most widely used in the construction industry include Primavera, MS Project, Powerproject, and so on. This section, using MS Project as an example, discusses how to build a project plan with the help of these programs. The discussion draws on the on-line documentations of the MS Project program.

6.2.1 Define a Project

At the beginning of project planning, it is necessary to define the objectives, assumptions, and constraints of the project. This needs to be done before starting the computer program. Objectives should be clear and measurable. They should include a list of deliverables, milestones or dates of delivery. At the initial stage, there are often many unknown factors, e.g., the resource requirement of key activities and the overall available resources at different stages of the project. Experienced planners can rely on their previous knowledge and come up with 'educated guesses'. The accuracy of these assumptions will determine the quality of the plan. Constraints on a project are factors that are likely to limit the project manager's options. Typically, the three major constraints are schedule, resources and scope. Any change in one of these constraints usually affects the other two, and also affects the quality of the overall plan.

Given the unknown factors and potential changes in constraints, the planner needs to define a scope management plan to deal with the inevitable changes to a plan during execution of a project.

Once the initial planning is completed, the planner can start a new file in the MS Project program. Figure 6.4 shows the user interface of the program.

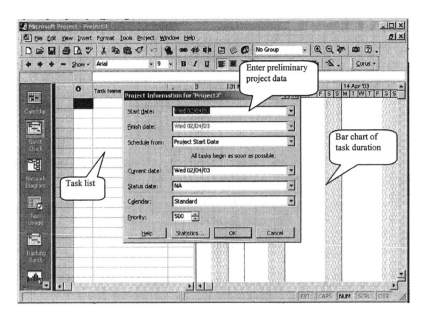

Figure 6.4 Start a new project schedule file

When an MS Project starts, it shows the Gantt chart view by default. The first task is to set the start and finish dates and some other general project data. Then it is ready to enter a list of individual tasks.

6.2.2 Enter a Task List

The basic element of a project schedule is a task. A task is a work package with a clear deliverable, a starting date and duration. MS Project provides a spreadsheet-like window, which enables tasks to be entered (Figure 6.5). If a task list already exists in another file, it can be copied or imported into a new schedule file.

Tasks should be entered in the order they will occur. For each task, enter its task name, duration and starting date. The program will calculate the ending date automatically.

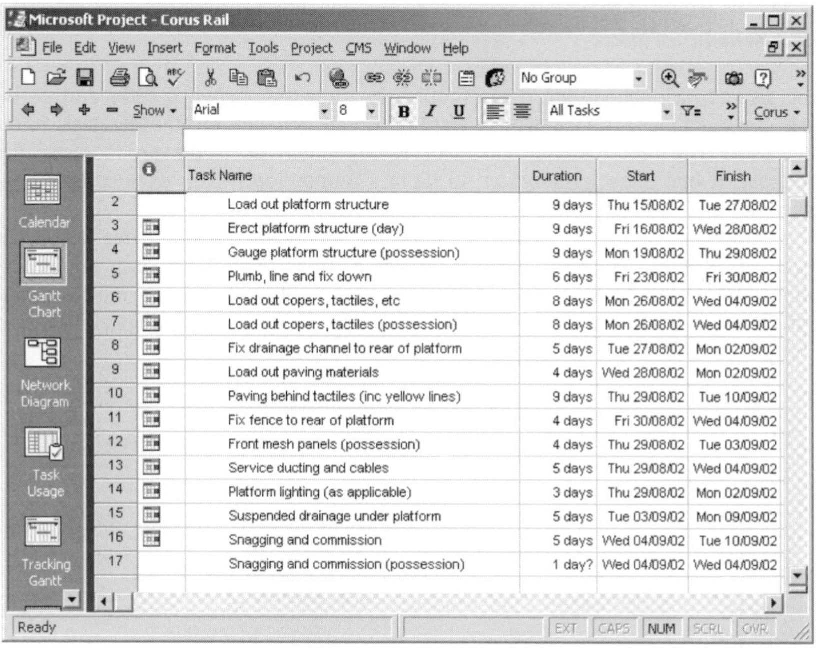
Microsoft Project - Corus Rail
File Edit View Insert Format Tools Project CMS Window Help
No Group
Show
Arial
8
All Tasks
Corus
Calendar
Gantt Chart
Network Diagram
Task Usage
Tracking Gantt
Task Name
Duration
Start
Finish
2 Load out platform structure 9 days Thu 15/08/02 Tue 27/08/02
3 Erect platform structure (day) 9 days Fri 16/08/02 Wed 28/08/02
4 Gauge platform structure (possession) 9 days Mon 19/08/02 Thu 29/08/02
5 Plumb, line and fix down 6 days Fri 23/08/02 Fri 30/08/02
6 Load out copers, tactiles, etc 8 days Mon 26/08/02 Wed 04/09/02
7 Load out copers, tactiles (possession) 8 days Mon 26/08/02 Wed 04/09/02
8 Fix drainage channel to rear of platform 5 days Tue 27/08/02 Mon 02/09/02
9 Load out paving materials 4 days Wed 28/08/02 Mon 02/09/02
10 Paving behind tactiles (inc yellow lines) 9 days Thu 29/08/02 Tue 10/09/02
11 Fix fence to rear of platform 4 days Fri 30/08/02 Wed 04/09/02
12 Front mesh panels (possession) 4 days Thu 29/08/02 Tue 03/09/02
13 Service ducting and cables 5 days Thu 29/08/02 Wed 04/09/02
14 Platform lighting (as applicable) 3 days Thu 29/08/02 Mon 02/09/02
15 Suspended drainage under platform 5 days Tue 03/09/02 Mon 09/09/02
16 Snagging and commission 5 days Wed 04/09/02 Tue 10/09/02
17 Snagging and commission (possession) 1 day? Wed 04/09/02 Wed 04/09/02
Ready
EXT CAPS NUM SCRL OVR

Figure 6.5 Task list window

Task Hierarchy

A complex project can have hundreds of tasks. These tasks can be grouped into different phases. A phase is a more manageable piece of work, which consists of a small collection of tasks. The phase itself is called a summary task and the tasks within a phase are called sub-tasks. Clearly, the start and finish dates of a phase are determined by the earliest start date and latest finish date of its sub-tasks. The sub-tasks can have their own lower level sub-tasks. The whole schedule becomes a task hierarchy with several levels.

Milestones
Milestones are intermediate goals that need to be achieved in order to realise the overall objective of a project. They are also review points where project progress can be assessed. In MS Project, a milestone can be set by entering a task with zero duration.

Recurring Tasks
Recurring tasks are tasks that repeat regularly, such as weekly meetings. A recurring task can take place daily, weekly, monthly, or yearly. It can be entered by specifying the duration of each occurrence, when it will occur, and for how long or how many times it should occur.

6.2.3 Schedule Tasks

The aim of task scheduling it to establish relationships between tasks and define task dependencies. A task whose start or finish depends on another task is the successor. The task that the successor is dependent on is the predecessor. For example, if one links 'excavation' to 'laying foundation', then 'excavation' is the successor and 'laying foundation' is the predecessor.

After the tasks are linked, changes to the predecessor's dates affect the successor's dates. Microsoft Project creates a finish-to-start task dependency by default. Because a finish-to-start dependency does not work in every situation, one can change the task link to start-to-start, finish-to-finish, or start-to-finish to model a project realistically.

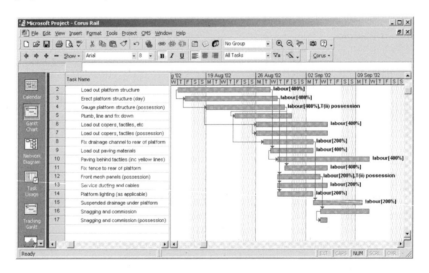

Figure 6.6 Schedule and task dependencies

Other scheduling functions include:

- *Overlap tasks or add lag time between them*: After sequencing tasks by linking them, the planner can overlap or delay them as well. In Microsoft Project, it is possible to delay tasks by adding a lag time to the predecessor task, and overlap tasks by entering a lead-time.
- *Set a specific start or finish date for a task*: Schedules can be made by entering task durations, creating dependencies between tasks, and then letting Microsoft Project calculate the start and finish dates. However, sometimes, the planner might want to set a specific start or finish date for a task. Task constraints that tie tasks to specific dates are called inflexible constraints; the most inflexible constraints are specific start or finish dates.
- *Add a deadline to a task*: When a deadline is set for a task, Microsoft Project displays an indicator if the task is scheduled to finish after the deadline.
- *Split a task into segments*: The planner can split a task as many times as necessary, if work on the task is interrupted and then resumes later in the schedule.

6.2.4 Assign Resources

Microsoft Project allows the user to create a list of the people, equipment, and material resources that make up a team and carry out the project tasks. The resource list consists of work resources and material resources. Work resources are people or equipment; material resources are consumable materials or supplies, such as concrete, wood, or nails.

Once a resource list is created, items in the list can be assigned to tasks. The user can assign more than one resource to a task and specify whether a work resource works full-time or part-time on a task. As more resources are assigned to a task, Microsoft Project automatically decreases the duration of that task. For example, a task with a one-day duration and one assigned resource has 8 hours of work. With effort-driven scheduling, if a second resource is assigned to it, the task still has 8 hours of work, but its duration is reduced to half a day. The program also gives the option to switch off effort-driven scheduling before assigning another resource. The task will then have 16 hours of work and still have a one-day duration.

Other resource related functions include:

- *Check and edit resource assignments*: The Resource Usage view shows project resources with their assigned tasks grouped underneath them. Using the Resource Usage view, you can find out how many hours each resource is scheduled to work on specific tasks and see which resources are over-allocated. You can also determine how much time each resource has available for additional work assignments.

- *Assign costs to resources*: Microsoft Project allows rates to be assigned to human and material resources to manage project costs accurately. Standard rates, overtime rates, or per-use rates can be assigned to resources.
- *Set fixed task costs*: When an exact cost associated with a task, such as equipment costs, is known a fixed cost can be entered.
- *Define when costs accrue*: In Microsoft Project, resource costs are prorated by default. Their accrual is distributed over its duration. It is possible, however, to change the accrual method so that resource costs take effect at the start or end of the task instead.
- *See the cost of tasks or resources*: After rates are assigned to resources or fixed costs to tasks, users may want to review the total cost of these assignments to make sure they fall within their expectations. If the total cost of a task or resource does not meet their budget, users may need to examine each individual task's costs and each resource's task assignments to see where costs can be reduced.
- *See the cost of the entire project*: Users can view their project's current, baseline, actual, and remaining, costs to see whether they are staying within the overall budget. These costs are updated each time Microsoft Project recalculates the project.

6.2.5 Optimise a Plan

A planner can optimise a schedule using a number of methods:

- *Check and adjust a task dependency*: A task dependency describes how a task is related to the start or finish of another task. Microsoft Project provides four task dependencies: finish-to-start (the most commonly used dependency), start-to-start, start-to-finish, and finish-to-finish. By using these dependencies effectively, a planner can modify the critical path and shorten project schedule.
- *Overlap tasks*: If some tasks can begin earlier than shown in the initial schedule, the planner can make them overlap (add lead time) to reflect more accurately how the work will be done. For example, if the electricians can begin wiring outlets before the walls are all finished, time can be used more efficiently by starting the 'Wire outlets' task after half of the walls have been roughed in. To do this, the planner sets up a lead time between the finish of the 'Rough-in walls' task and the start of the 'Wire outlets' task.
- *Check and adjust constraints on tasks*: Task constraints can be used to create a more accurate schedule by tying tasks to specific dates. For example, the planner can specify that a task must start no earlier than a particular date or finish no later than a particular date.
- *Make tasks shorter by adding more resources*: If a project has more flexibility with resource assignments than schedule deadlines, adding resources can be an effective way to shorten its project schedule.

6.3 TRACK AND MANAGE A PROJECT

To manage a project, a project manager needs to monitor the elements of the project triangle: time, money, and scope. Adjusting one of these elements affects the other two. Events such as unexpected delays, cost overruns, and resource changes can cause problems in the project schedule. Using MS Project, a manager can keep project information up to date and identify problems early.

6.3.1 Manage the Schedule

Once work has begun on a project, its planned schedule can be used to keep track of actual start and finish dates, tasks' percentage of completion, and actual work. This is achieved through a list of functions.

- *Check if tasks are progressing according to plan*: To keep a project on schedule, make sure that tasks start and finish on schedule. The Tracking Gantt view helps find trouble spots, tasks that vary from the baseline plan. The manager can then adjust task dependencies, reassign resources, or delete some tasks to meet deadlines. The Tracking Gantt view pairs the current schedule with the original schedule for each task.
- *Enter actual start and finish dates for a task*: Tasks that start or finish late can throw an entire project off schedule by delaying the start or finish dates of related tasks. Tasks that start or finish early can free resources to work on other tasks that are behind schedule. Microsoft Project uses the actual values entered by the user to reschedule the remaining portions of a project.
- *See if tasks have more or less work than planned*: To manage resource assignments, a project manager needs to make sure resources complete tasks in the time scheduled. If a project baseline is saved, the manager can check the variance information. Variances in a schedule can be good as well as bad, depending on the type and severity of the variance. A task with less work than planned, for example, is usually good but may indicate that the resources are not allocated efficiently.
- *Compare actual task information to the baseline*: When a baseline plan is updated with actual operation information, a project manager can compare the baseline plan to the actual progress to identify variances. Variances alert the manager to the areas of the project that are not going as planned. Every project has variances, but it is important to find tasks that vary from the baseline plan as soon as possible so that task dependencies can be adjusted, resources reassigned, or some tasks deleted to meet the deadlines.

6.3.2 Manage Work

A project manager may need to track how much work each resource on a project completes task by task or cumulatively for the project. By doing so, the manager can compare the planned and actual amounts of work. This comparison can help to keep track of project resources' performance and plan workloads for future projects. MS Project allows users to:

- *Enter the total actual work done by a resource*: If a schedule is produced based on the availability of resources, tracking the progress of tasks by updating the work completed on a task can help to determine how each resource is performing.
- *Update a resource's actual work by time period*: The user can track actual work for individual resources using the time-phased fields in Microsoft Project. Tracking resources' actual work by using the time phased fields can help to keep a project up to date by time period, because the user can enter information for a particular day (or other time period) in a schedule.
- *See the variance between a resource's planned and actual work*: The users can analyse how much total work a resource is accomplishing by looking at the variance between the baseline work and actual work. They can also compare those figures to the baseline work and actual work over time, to see how the resource's work is progressing in greater detail.

6.3.3 Track Costs

A project manager needs to track cost overruns in a phase of a project or learn how much a particular resource costs on a certain day. Tracking costs can help a manager see where changes need to be made to finish a project on time and within budget. MS Project offers the following support:

- *Enter actual task costs manually*: Microsoft Project automatically updates actual costs as a task progresses based on the task's accrual method and the rates of the resources. But to track actual costs separately from the actual work on a task, costs can be entered manually instead. To update costs manually the automatic updating of actual costs must be turned off and then the actual cost for an assignment after the remaining work is zero must be entered.
- *Update actual costs by time period*: Actual costs can be tracked using the time phased fields in Microsoft Project. Tracking actual costs using the time phased fields can help keep the project up to date by time period because information can be entered for a particular day or other time period in the schedule.
- *See if tasks cost more or less than budgeted*: If fixed costs are assigned to tasks or wages specified for resources, the user may want to see tasks that cost more than budgeted. By creating a budget using a baseline plan and closely tracking project costs, it is possible to catch cost overruns early and adjust either the schedule or budget accordingly. Microsoft Project calculates the cost of each resource's work, the total cost for each task and resource, and the total project cost. These costs are considered scheduled or projected costs, which reflect the latest cost picture as the project progresses.

- *See the total project costs*: It is possible to view the project's current, baseline, actual, and remaining costs to see whether your budget is being kept to. These costs are updated each time Microsoft Project recalculates the project.
- *Analyse costs with the Earned Value table*: When wanting to compare the expected progress with the actual progress to date, it is possible to use the Earned Value table. It compares, in terms of costs, each task's baseline schedule with the actual schedule. It is also possible to use the Earned Value table to forecast whether the task will finish under or over budget based on the cost incurred while the task is in progress. For example, if a task is 50% complete and the actual cost incurred to date is £200, it is possible to see if £200 is more than, less than, or equal to 50% of the baseline (or budgeted) cost. The VAC field displays the variance at completion between baseline cost and scheduled cost for a task.

6.3.4 Balance Workload

The user should check your schedule for resources with too much or too little work. If some resources are over-allocated, see if adding more resources to a task or reassigning a task will give the results wanted. If this does not work, it is possible to delay tasks assigned to an overworked resource until later in the schedule or reduce the amount of work for tasks.

- *Find overallocated resources and their task assignments*: People and equipment are over-allocated when they are assigned more work than they can complete in their scheduled working hours. Before resolving over-allocations, it is necessary to determine which resources are over-allocated, when they are over-allocated, and what tasks they are assigned to at those times. To resolve the problem, the people and equipment must be allocated differently or the task must be rescheduled to a time when the resource is available.
- *Reduce a resource's work*: After assigning a resource to a task, it is possible to change the total work values for the resource's work on the task or change work values for a specific time period when the resource works on the task. Tailoring work values this way can make a schedule more accurate at a finer level of detail.
- *Reassign work to another resource*: If trying to resolve a resource over-allocation using other methods and the over-allocation persists, it may be time to reassign the task to another resource with more time. This is an alternative method of manually levelling a schedule by reassigning work rather than delaying work.
- *Delay a task*: A simple way to resolve a resource over-allocation is to delay a task assigned to the resource until the resource has time to work on it. A delay can be added to a task, the effect on the resource's allocation checked, and then the delay further adjusted if necessary. Delaying a task also delays the start dates of its successors and can affect the finish date of the schedule. To avoid this, delay tasks with free slack

first (non-critical tasks) and only delay them up to the amount of slack that is available for each task. Experiment with adding delay to different tasks to see the effect on the schedule.

- *Change a resource's working days and hours*: The project calendar designates the default work schedule for the project, but it is possible to create a resource calendar to indicate work hours, vacations, leaves of absence, and sick time for individual resources.

6.4 SITE MANAGEMENT

6.4.1 Operation Simulation

In Chapter 4 we saw the importance of simulation to represent the physical processes occurring within the built environment. Simulation is also useful for on-site construction operations that require complex co-ordination between a number of activities, such as material delivery, equipment movement and building work. An efficient operation depends on good synchronisation of all these activities.

Case study
The Virtual Site Management Tool

A novel application of remote controlled audio/visual technology has demonstrated reductions in defects and improved predictability. The system is marketed as Virtual Site Management (VSM).

VSM was developed and trialled in The Paddock project, a six-dwelling, £450k private development by L Green (Kirkby) Ltd, completed in August 2001. With VSM it is possible to make inspections and surveys without the inspector setting foot on the site. A semi-skilled operator can be directed from any remote location to perform condition surveys and progress inspections, all with customised control at every stage.

VSM uses either fixed dome type cameras for general site monitoring, or portable, hand-held devices for investigation of specific problems or features. Images are compressed and transmitted with audio over an ISDN line or Internet connection. Managing director David Green concedes that trade contractors and operatives were initially suspicious of the 'big brother' implications. 'The key is making the system accessible to everyone,' he says. 'It's a 2-way thing. They strike a problem and can get it resolved quickly. We don't need to visit the site in most cases.'

Dale Evans of Dukeries Tiling, one of the trade contractors involved in the trial, used VSM on the project's dedicated web site. He says: 'There's a fair potential for communicating with the site, solving problems without having to be there.'

Source: IT Construction Best Practice

---

computer Operation Simulation system, such as SIMUL8 Planner (www.simul8.com), emulates what happens at a construction site in the real world by representing the workers, machines and materials, and computing the cycle times of each step, taking many uncertain factors into consideration. Its aim is to identify production bottlenecks, reduce under-utilised resources and develop a more productive operation sequence.

6.4.2 Site Monitoring

On a typical construction site, the main contractor usually employs a number of sub-contractors to undertake a variety of tasks. There is a need for a site monitoring system, firstly, to ensure that only authorised personnel are allowed on site, secondly, to keep track of the time spent on site by each subcontractor. An example is the Construction Site Monitor (CSM) developed by Public Access Terminals (PAT) Ltd (www.girovend-pat.co.uk/pat/). The PAT's Construction Site Monitor system integrates swipe card technology and a computer program. It produces an ID card for each individual member of the subcontracting team. The card bears the holder's photograph, and incorporates a magnetic stripe or bar code for access to and from the site. Using this system, the access time to a site can be set for each individual. The authorised users can have unhindered access within designated working hours but not at any other time. This means that subcontractors will be on site only when they are needed.

The system provides a real-time log of all entry and exit data from access points throughout the site. This log can be used to provide a report on which individual from which sub-contracting team was on site at a particular time, as well as being able to aggregate the total time spent on site during any specified period for payment purpose. The information can also be used to check compliance with the EU and Central Government Directives on work hours and local labour usage.

6.4.3 Material Procurement

Procurement for contractors at the on-site stage involves obtaining suitable materials and subcontractors in time to keep the field construction moving according to plan. A typical sequence of steps for the procurement activity was outlined by Paulson (1995):

1. Recognise the need and issue requisitions
2. Prepare or adapt specifications
3. Advertise for bids or solicit price quotations
4. Receive and evaluate bids or quotations
5. Issue purchase orders or subcontracts
6. Prepare shop drawings or samples
7. Approve shop drawings or samples
8. Fabricate
9. Ship

10. Deliver and unload
11. Inspect and accept or reject
12. Store
13. Use item on the project.

The key to an IT based procurement system is the use of a database. A purpose built database system provides a central repository of information on the need for materials, equipment and services at every phase of the construction process. The scheduling function of such a system will prompt the contractors to take procurement actions at the appropriate time and to manage schedule changes as a result of delays by the suppliers or subcontractors.

6.5 MOBILE COMPUTING AND ITS ON-SITE APPLICATION

6.5.1 Wireless Communication

Unlike an office situation, on construction sites people are more mobile. To carry out their job, communication with others is essential and quality, quantity and timing of information can either hinder or facilitate successful results. At present, most sites do not have access to fixed line computer networks. Setting up a dedicated network for the duration of a project is usually an expensive option, which is only suitable for large projects. The latest advances in wireless technologies offer a potential solution to the on-site communication problem.

For many years, workers on construction sites used walkie-talkie to talk to each other and coordinate activities. Lately mobile phones have become a more flexible alternative. However, until now, these devices have only been good at voice communication. They are not yet suitable for data exchange.

The engineering design consultant company, ARUP, has carried out a study on 'Construction Site Communication' for the IT Best Practice Programme in the UK. The study suggested that wireless Local Area Networks (LANs) can be set up in a number of ways. The first and most simple of these, is an ad-hoc network, where a group of PCs and mobile computers with wireless LAN cards in them can communicate with each other directly. This allows the transfer of information between PCs although it does not allow access to a fixed network. The most popular method of connectivity for wireless LANs is as an extension to a wired network allowing mobile users or other authorised visitors access to the main servers and other work related information.

In addition to desktop and laptop computers, there are numerous new, and more portable, devices that can be linked up to the wireless LANs:

- *Personal Digital Assistants (PDAs)*: A PDA, also known as a Palmtop computer, is a mobile computer that fits into the palm of a hand. It allows storage, access, and organisation of information. Basic Palmtops allow storage and retrieval of addresses and phone numbers, maintain a calendar, and create to-do lists and notes. More sophisticated PDAs can run word processing, spreadsheet and industry specific applications and also provide e-mail and Internet access.

- *Hand-Held Computers*: They offer the main functionality of a laptop in a smaller package. They feature a keyboard and a landscape display. Typically they will run on a Windows based system, and operate in a similar way to a desktop PC. Due to the keyboard these are probably best suited to use where there is a stable platform available, for example in a pick-up truck.
- *Pen Tablet / Touch Computers*: A Pen Tablet is a computer that runs a full desktop operating system but utilises an electronic pen (called a stylus) rather than a keyboard for input. Pen computers, like Palmtops, generally require special operating systems that support handwriting recognition so that users can write on the screen or on a tablet instead of typing on a keyboard. They are often 'docked' when in the office enabling connection to a desktop PC.

6.5.2 Applications On-Site

The following is a list of examples of wireless computing applications that assist on-site processes:

- *CAD Applications*: CAD is now used on almost all construction projects to produce drawings for use in the field. However, although the drawings are produced electronically, they are printed out for use. This eliminates many of the advantages of electronic production, and reduces the opportunities for effective feedback.
- *Collaboration Software*: currently there is much discussion about web based collaboration systems solving the industries' fragmentation problems. However, these systems are not yet in common use in the field e.g. by foremen and site engineers. The flow of electronic information therefore comes to an abrupt halt when it reaches the construction site and thus many of the efficiency and knowledge-based benefits are lost. Collaboration Software suppliers are beginning to address this issue by extending key collaboration features to mobile users in the field either through their mobile phones or other handheld devices.
- *Data Capture*: In this area it is possible to choose between purchasing tools for specific applications e.g. Time Sheets or Inspections, and purchasing form-building software or services to create business applications for mobile computing. They are typically used for inspection and checking tasks where, traditionally, information needs to be recorded using pen and paper.
- *Project Management*: The project administration area overlaps with some of the features that collaboration tools offer. However, there are also software applications available that add project and programme management capabilities.

The above information comes from the ARUP study mentioned. The ideas are by no means exhaustive. This is a rapidly developing area of great potential. New types of applications are emerging all the time.

6.6 ON-LINE RESOURCES

Organisations

http://www.pmi.org
The Project Management Institute (PMI) is a leading non-profit professional association in the area of project management. PMI establishes project management standards, provides seminars, educational programs and professional certification that more and more organisations desire for their project leaders.

Scheduling and Planning Software

http://www.primavera.com/
Primavera is a leading supplier of project planning software. It provides a family of related products including: Primavera Enterprise – enterprise-wide project management; Primavera Project Planner – multi-project, multi-user project and resource planning and scheduling; Primavera Expedition – complete multi-project controls and contract management; PrimeContract – project collaboration and construction process automation; SureTrak Project Manager – resource planning and control for small-to-medium sized projects; and so on.

http://www.constructasoftware.com/site_management.htm
Constructa provides site managers with access to information about deliveries, suppliers, contracts, personnel, H&S, etc. Receipt of goods can be recorded electronically while on site to be synchronised with HQ.

http://www.microsurvey.com/products/mscad/index.htm
MicroSurvey CAD 2002 is a complete desktop survey and design program specifically for surveyors, contractors, and engineers. No plug-ins or modules are necessary.

http://www.kidasa.com/
Milestone Professional is an easy to use project management and scheduling software. It can be used to create Gantt, timeline, and milestones charts. It can also produce presentation report for MS Project schedules.

http://www.ccssa.com/
The Construction Computer System Planning system is a simple yet powerful critical path networking system that has been designed for the contractor. It can be used to draw a bar-chart or to maintain a detailed precedence network, or as a mix of these two methods. The network can be entered in precedence fashion or as a linked bar-chart that will automatically construct the precedence network.

http://www.amsrealtime.com/products/projects.htm
AMS REALTIME Projects is a powerful and easy-to-use tool for integrated project management, planning, scheduling and cost management. It supports the

needs of individual project managers and provides consolidation, aggregation, analysis and management via powerful multi-project facilities in standalone mode or across the enterprise.

http://www.astadev.com/
Asta Teamplan is a mid-market, enterprise-level programme and resource management software application for IT and services organisations. A proven, scalable tool, it enables organisations to realise their strategic objectives through the effective management of people, costs, income and deliverables across multiple projects and ongoing work. PowerProject TeamPlan is a very flexible project and programme management solution with its menu structure conforming to the Microsoft Office model. It has an excellent, fully-featured Gantt chart.

http://www.focus5.co.uk/
This is an add-on to Microsoft Project. It supports the principles of the Theory of Constraints.

http://www.omnis-group.net/
Integrated Project Support Office (IPSO) is a project and programme management tool. It allows users to produce plans and maintain plans. It integrates project and time management processes and controls links to document with tasks.

http://www.e-pso.co.uk
e-PSO is primarily a process management solution that will provide organisations with control and visibility of their entire project portfolio. It is a unique product in its use of workflow technology and focus on processes.

http://www.planview.com
PlanView is an integrated web-native solution helps organisations to manage portfolios, projects and resources. It is a programme and project scheduling, management and reporting tool. PlanView is particularly strong in the resourcing area and has a good timesheet system.

http://www.clarityds.com/
Project Organiser is an integrated solution to documentation control, activity management and time recording. It helps a company to enforce its standards and methodology in the same system as that used for recording time spent on projects, and tracking all activity and issues.

http://www.coco.co.uk/prodpks.html
Project KickStart is an easy to use planning and scheduling program.

On-Site Communications and Management

http://www.roke.co.uk/markets/industry_automation/construction.asp
Roke Manor Research provides communications solutions for on-site construction.

http://www.cordis.lu/infowin/acts/rus/projects/ac088.htm
Mobile Integrated Communication in Construction (MICC) was a EU research project that aimed to introduce the use of on-site mobile communications as a way of improving the global competitiveness of the European construction sector.

http://www.pocketcad.com/
PocketCAD specialises in mobile devices for site measurement and survey data collection.

http://www.autodesk.com/
Autodesk OnSite can read spatial data from different sources and present them in one display. It can be used to produce as-built models. Autodesk Survey is an AutoCAD-based solution for the survey, land planning, and engineering industry. By interfacing with a host of industry-standard survey instruments, Autodesk Survey automatically creates linework, points, symbols, and terrain-model break lines directly from field survey data.

http://www.pocketpccreations.com
Pocket PC Create provides a range of PDA-based applications that can be used for on-site inspection purposes.

http://www.esri.com/
ESRI is a leading GIS solutions provider. Many of its software products can be used for surveying and on-site data capturing.

http://www.onsyss.com
Onsyss provides planning and scheduling applications for PDA platforms. These systems, include OnSite, OnTrak and Primavera Mobile Manager, can be used for project management at on-site operation stage.

PM Portals

http://www.pmtoday.co.uk/
Project Manager Today is a portal for project management related resources.

http://pmri.net/
This site provides information on state-of-the-art productivity tools, consulting and training for enterprise wide project management implementation.

SUMMARY

This chapter introduced several types of IT system used during the on-site construction phase. Planning and scheduling systems help to control and monitor the operations and resources of a construction project. Procurement systems assist in buying the right materials and services at the right time. Surveying systems automate some tasks traditionally carried out manually by surveyors. On-site operation simulation systems can be used to emulate and evaluate various options to establish the most efficient operational arrangements.

DISCUSSION QUESTIONS

1. Consider a project with which you have been involved. What planning tools were used for scheduling this project? Do you think any of project planning tools would have made it easier to plan and control the work?
2. What advantages does computer based surveying offer over traditional techniques?
3. The principles of simulation have been established for many years. Why has it been little used until recently? What alternative 'rules of thumb' are commonly used in the construction industry?

CHAPTER 7

Computer Aided Facilities Management

LEARNING OBJECTIVES

1. Understand the process of facilities management.
2. Appreciate the potential of computers for facilities management.
3. Gain knowledge of the main types of computer aided facilities management system.
4. Discuss ideas about the future of intelligent buildings.

INTRODUCTION

Faults in building fabric and facilities cause serious disruption to normal business for bank branches, retail shops and similar buildings in the service sector. For example, if a shop entrance door cannot open customers will not be able to enter and buy goods. Similarly, if a refrigerator has broken down in a food store on a hot summer day, food will be ruined. Therefore, building and facilities maintenance is essential to ensure a healthy and safe environment for business activities. In the last few decades, the importance of building repair and maintenance has been steadily increasing. In the late 60s, this type of work represented 28% of the total construction output (Seeley, 1976). The figure rose to 38% in 1979, 46.4% in 1985 and accounted for over 50% of construction industry output in 2000 (DETR, 2000).

Facilities Management (FM), which originated in the USA, is the function that ensures a building and its services are in good order and they accommodate the occupants to their optimum advantage. It is not only about maintenance and repairs but also about asset management and space planning when the use of a building changes over time.

Buildings are important because they are often the biggest fixed assets of a company and they provide the physical environment in which normal business can be conducted. During use, defects can occur in buildings for a variety of reasons, such as those caused by bad design, exposure to weather, damage due to wear and tear, etc. When faults occur, they need to be repaired to prevent further damage. The built environment consists of the building fabric itself and the many service systems, such as lighting, heating, ventilation and security. These service systems also need to be maintained. Furthermore, in today's business world, constant change is a common feature. When business activities are reorganised, the requirement for space and service support usually changes as well.

FM is a complex process involving many tasks. Some of the tasks may be outsourced to external contractors and subcontractors. Effective coordination and information flow are essential. Computers are playing an increasingly important role in the FM process.

This Chapter will introduce the main functions of existing computer aided FM applications. It will also discuss future development of advance building control systems and the ultimate goal of the *intelligent building* that can respond automatically to its occupants' needs.

7.1 THE ROLE OF FACILITIES MANAGEMENT

7.1.1 Core and Non-core Business Activities

All organisations have core and non-core business activities. The core business is the customer interface or the point of delivery of the product and services. For example, for a high street bank this could be the local branch or the enquiry desk at head office. In the case of a car supplier, the point of contact is the garage forecourt. The role of the non-core, or supporting, activities is to create the environment in which core business can be delivered efficiently and cost effectively. Non-core business is by no means unimportant. In fact supporting services are crucial in avoiding disruptions to production and product delivery. The non-core activities of a typical organisation include administration, IT, finance and FM. FM is responsible for strategic planning, space planning, estate management, lease management, asset management, etc.

7.1.2 Growth of FM

Facilities management can mean many things and it may be applied in different ways by different companies. In some cases, it might refer to the out-sourcing of IT services. In other cases it could refer to the management of the mechanical and electrical services within a building. The emergence of FM can be traced back to the late 1960s. Since then, it has gradually gained a foothold as a discipline and profession within the property and construction industry (Tay, 2001). The status of FM as a separate profession is helped by the establishment of several related profession bodies, including the British Institute of Facilities Management (BIFM) and the International Facility Management Association (IFMA). However, it is still a relatively new profession and the exact scope is still being debated. This debate is reflected by a variety of definitions of FM offered by different authors:

> The definition from the American Library of Congress is *"the practice of co-ordinating the physical work place with people and work of the organisation integrates the principles of business administration, architecture and the behavioural and engineering sciences."* (Spedding, 1994)
>
> *"FM is responsible for co-ordinating all efforts related to planning, designing and managing buildings and their systems, equipment and*

furniture to enhance the organisation's ability to compete successfully in a rapidly changing world." (Becker, 1990)

"The practice of co-ordinating the physical workplace with the people and work of an organisation; integrates the principles of business administration, architecture, and the behavioural and engineering sciences" (NHS Estates, 1996)

"The practice of FM is concerned with the delivery of the enabling workplace environment, the optimum functional space that supports the business processes and human resources" (Then, 1999)

Despite some differences of interpretation, there is also strong agreement. FM covers both 'hard' services such as maintenance of fabric and engineering plant, and 'soft' services such as cleaning, security and catering (Armstrong, 2002). In the context of this book, the focus is on the 'hard' services aspect of FM.

7.2 COMPUTER AIDED FM

Computer Aided FM (CAFM) is about the use of computer programs to automate certain FM tasks. The computing technologies that underlie CAFM include databases, CAD, Graphic Information Systems (GIS), computer networks, etc. CAFM systems can be broadly divided into three categories:

1. Asset management systems. Traditionally asset inventories were logged on paper. It was difficult to keep them updated and difficult to extract data from them for the purpose of analysis. The emergence of spreadsheet and database technology has provided an opportunity to develop asset management systems. Using such a system, a large amount of data about assets, such as space and equipment, can be entered in structured formats. The data can then be extracted easily and used for maintenance scheduling and analysis of tasks.
2. Graphical spatial planning systems. FM functions cover spatial planning and positioning of equipment and furniture. CAFM systems, based on CAD, GIS and 3D technologies, enable the facilities managers to carry out these tasks with the assistance of graphic displays. Many of these systems also integrate seamlessly with asset management databases.
3. Computer integrated facilities management systems. The purpose of these systems is to integrate distributed FM activities. The development of computing networks, especially the Internet, has made this possible.

At present, there are many 'off-the-shelf' FM programs on the market. Their typical functions are discussed in the following.

7.2.1 Helpdesk

Many companies manage their FM work, in particular maintenance work, through help desk call centres. The process involves several parties: facilities users, call centre operators, contractors, and FM managers. The facilities users are those people who use the building and its facilities to carry out business activities. When a fault occurs, the user will contact the helpdesk to report the fault. Helpdesk operators are responsible for logging the report and recording details of the fault. Contractors are responsible for carrying out the repair work. The FM managers are responsible for making decisions on contracts and whether a job needs to be carried out. Figure 7.1 illustrates the process of a typical maintenance job.

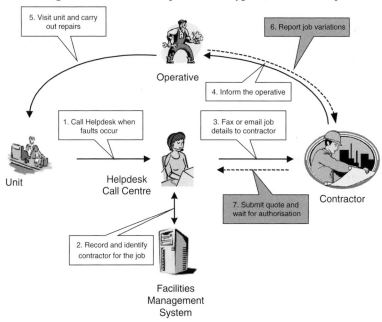

Figure 7.1 Maintenance job process using helpdesk system

1. A job starts when a facilities user in a business unit reports a fault to the FM helpdesk by telephone, fax or e-mail.
2. The helpdesk operator, using a standalone FM system, records information about the unit, the problem that has occurred and allocates the job a reference number. The system identifies a contractor according to the type of building works required and the unit's location.
3. The helpdesk then prints out the job description and sends it to the contractor by fax. Sometimes the information is sent using e-mail.

4. Based on the job description sent by the helpdesk, the contractor assigns an appropriate operative (in-house engineer or subcontractor) to the problem.
5. The first thing that the operative does is to estimate the cost of fixing the problem.
6. If it is over a certain budget limit the operative will provide the contractor with an estimated cost for authorisation purposes. Otherwise, there is a check on whether any equipment or spare parts are needed from a supplier and then the repair work can be started.
7. For repairs over the budget limit, the contractor needs to request authorisation from the FM manager before the work can be carried out.

The helpdesk function is to enable the operators to efficiently log and monitor reported faults. For every job a lot of information needs to be recorded, including: location, and access details, nature of the fault, asset identity, etc. Most FM programs have user-friendly interfaces to facilitate the capture of these data.

After entering the input data, helpdesk programs are able to identify suitable contractors according to criteria such as skill requirement, geographic location and level of service agreement. The job details can be printed out and faxed to the contractor. Some new programs utilise e-mail and the Internet to send the information to the contractor very quickly.

7.2.2 Asset Management

Assets here refer to equipment and furniture owned by an organisation. There are many reasons to track assets within an organisation. For example, when a piece of equipment has broken down, it is necessary to know whether it is under warranty. An asset management system is, first of all, an asset register of 'what, where and which one'. In addition to location, standard type and unique ID, other information recorded might include purchasing date, leasing agreement, service agreement, depreciation, etc.

Today a typical organisation would have thousands, tens of thousands, or even more assets in the form of computers, printers, photocopiers, furniture, etc. Recording their details and keeping the information updated is a huge task. An effective method of doing this is to use bar coding. Although bar coding assets will not eliminate the need to enter the detailed information in the first place, it will make the tracking task a lot easier. Some asset management programs come with pre-loaded data for typical assets. When entering an asset the user can select a standard item then edit some of its fields. This prevents the painstaking task of typing everything from scratch.

7.2.3 Space Management

The aim of space management is to track the use of building spaces, e.g., how a space is used and who is using it. Space is an expensive asset. It is important to seek efficient space usage so that the occupancy cost per square metre can be reduced. A space management system can integrate intelligent databases and drawings to track the use of space in all the buildings of an organisation (Figure 7.2).

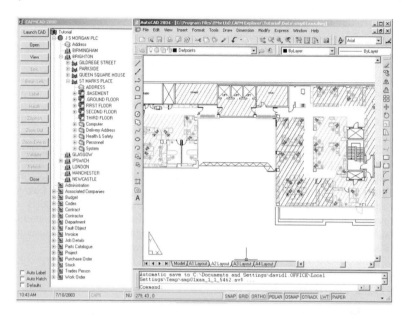

Figure 7.2 Space management system (image courtesy of www.cafmexplorer.com)

Space management systems, such as ArchiBus/FM, provide flexible methods for collecting and organising spatial information to support specific reporting requirements. They also provide room booking functions that allow the user to schedule the use of shared rooms or shared spaces based on availability, charge back rates, amenities, and seating capacity.

Space management systems can also help in space planning and move management. They have many tools that make it easy to show why more space, or a reassignment of space usage, is necessary. Simple space planning or moving can be achieved with a basic CAD system. However, if the activity involves complex relocation of people, furniture and equipment in multiple or multi-storey buildings, using a CAD system alone would be difficult. In addition, space management systems can provide forecasts of space needs based on headcount, functional usage, and logistics. This helps strategic decision-making at the organisation level and better understanding of the influences on occupancy costs at the operational level.

7.2.4 Building Operation Management

This is about maintenance management. Maintenance includes planned and reactive maintenance activities. Planned maintenance or predictable maintenance, also described as routine maintenance (CIOB, 1982), is the action taken to avoid expected or avoidable failure (Seeley, 1976). Unplanned maintenance, also described as reactive maintenance, is the type of work resulting from unforeseen damage or failure due to external causes or failures of planned maintenance. It is usually handled by a helpdesk (section 7.2.1).

Planned maintenance systems are management tools for forecasting workload relating to assets, property, or scheduled services. First of all they can help FM managers to schedule planned maintenance work. Through the system diary, they allow FM managers to view either monthly or annually forthcoming activities such as, planned maintenance, health and safety schedules, contract renewal and warranty expiry dates. They can also be used to manage work volume by providing status reports on the progress of multiple work orders from start to finish.

These systems help to reduce the maintenance costs and/or prolong the life cycles of equipment by helping to choose optimum maintenance strategies. By tracking accumulated repair and maintenance costs, FM managers can determine whether it is more cost-effective to repair or replace aging equipment. Learning from previous projects is also important in reducing maintenance costs and improving quality. For example, if a particular fault keeps occurring, preventive measures may be considered. Building operation management systems maintain an easily accessible maintenance history that can be reviewed and analysed.

7.2.5 Other Functions

CAFM systems provide a number of functions, such as real estate and lease management, master planning, overlay CAD with design management.

7.3 ON-LINE HELPDESK SYSTEM

At present, most large client organisations and FM consultant firms in the UK use stand alone helpdesk systems at their FM call centres, as described in section 7.2.1. These organisations often deal with a large number of reactive maintenance jobs annually, many of which are repetitive. The process of dealing with each job is relatively slow, and the administrative cost is high in comparison to the cost of a typical job itself. This section explains how improvement can be achieved using knowledge management and Internet communication technology.

7.3.1 Deficiencies of Stand Alone Helpdesk Systems

Inadequate Knowledge Support
One of the main functions of the helpdesk IT system is to help to identify the right contractor for a job. The helpdesk operators are responsible for choosing the type of work from a list on the system when someone reports a fault. It is important that the type of work entered into the IT system matches exactly what the unit needs, because it is the key criterion for choosing contractors. If the proper questions are

not asked, misunderstanding of the problem can occur which will lead to the wrong repair type being selected, and the wrong contractor being sent. Furthermore, if an inadequate job description is recorded, the contractor could send out a wrongly skilled or equipped operative.

Very often, novice helpdesk operators have no technical background and little knowledge about maintenance work. They are given a standard script for handling calls from the unit managers. They gain experience through:

- Training given by the company in charge of the helpdesk
- Day by day experience
- Other colleagues.

Once they have gained some experience, they may no longer need the standard script. The helpdesk operators build up tacit knowledge over time with their experience and knowledge gained from the job. However, because this is not explicitly captured and codified, this knowledge is lost to the organisation when an operator leaves. For a new operator to become similarly skilled will require either lengthy working experience or costly training. If the knowledge can be captured and disseminated, it will help in the training of new, as well as experienced, helpdesk operators.

The over reliance on human interaction between the unit managers and the helpdesk operators also requires a large number of operators to be employed at call centres. Knowledge Management techniques can be used to automate some of the interaction and reduce the need to speak to a human operator when a fault is reported. The clients involved in one study identified this as a potential cost saving area.

Double Handling of Data Entry

When job information is produced by the helpdesk system and sent to a contractor using fax or e-mail, the data are re-entered into the contractor's own IT system manually. Unfortunately, if there is no compatibility between the two systems, this can cause inefficiencies and, potentially, errors. Due to the lack of widely recognised data standards for the building maintenance domain, it makes integration between different maintenance systems particularly challenging.

Poor Communication Medium

Due to the lack of integration, paper is still widely used for data transfer between clients and contractors in the forms of faxes, memo messages, quotations, forms, reports and certificates. The parts of the process that still use paper are:

- Job description – the helpdesk prints out a copy of the job details and description from the system before it is sent to the contractor.
- Quotation – the contractor sends the quotation for a job to the facilities manager for approval.
- Worksheet and Feedback Form – the operatives are required to fill in a form on the work undertaken, and then get the assessment part of the form completed and signed by the unit.

- Certificate of Payment – the facilities manager certifies payment to the contractors upon approval of invoices, and then sends a certificate to the contractor.

The facsimile machine is the most frequently used device for the transfer of information. The Internet was not widely used by the companies included in a survey, especially the contractors. Besides the facsimile machine, some paper documents are also sent via post or handed in to the office by the operative once a week. The use of paper in this way causes delays in the entry of recent job details, as users must wait for the paper forms' arrival before the information can be entered into the system. Therefore, the status of current jobs cannot be checked on the computer, if queries are made from the unit. In addition, the job authorisation procedure involves a long communication chain from the operative – to the contractor's regional branch – to the contractor's helpdesk – to the client's helpdesk – to the client's managing agent and then all the way back again. Where approval is required, operatives have to revisit the unit after the arrival of the managing agent's approval, instead of carrying out the repair instantly.

The use of paper slows down the speed in which reactive maintenance work is handled. It also creates management difficulties. The filing of paper documents is space consuming, and certain levels of information recorded on the paper documents are abandoned. For example, the unit manager's comments in the feedback part of the worksheet are seldom used to assess the contractor's performance for payment approval, because they are not entered into any of the IT systems. The forms are filed and forgotten about, unless a complaint is made.

7.3.2 On-line Systems

In the last few years, there have been numerous efforts to explore the use of the Internet to assist maintenance work (Clark, 1997; Finch, 2000; So, 1997; Wang, 2002). Ali and Sun (2002) presented a knowledge-based on-line helpdesk system (Figure 7.3). The system was aimed at improving the communication between all parties involved in the process, and fault diagnosis. The system adopted a client and server configuration and web-based interface. It consists of a server, which runs both as a web server and a database server, and many client computers located at various clients' premises. The connection between the client and the server can be achieved through a modem via an Internet Service Provider (ISP), The Integrated Services Digital Network (ISDN – system that allows data to be transmitted simultaneously using end-to-end digital connectivity), or Asymmetric Digital Subscriber Line (ADSL – a continuously available connection for transmitting digital information at a high bandwidth). For some properties, such as public houses, where installing a computer is not a practical option, the unit manager can still contact the helpdesk through the telephone. The link between the server and the contractors requires a high level of security because quotations and payments will be transferred on-line. In such instances, a Virtual Private Network (VPN) will be installed for security purposes. Users interact with the server through a web-based interface.

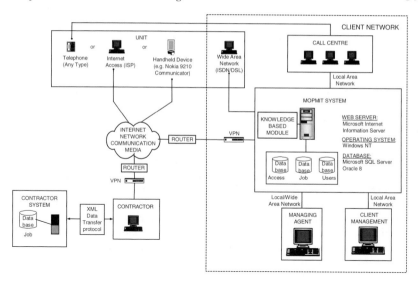

Figure 7.3 An on-line helpdesk system

The following scenario illustrates the new process of managing reactive maintenance work using the on-line system (Figure 7.4):

1. An electronic shutter at a retail shop has a fault and requires urgent repairs. The shop manager logs onto the Web server using a pre-allocated username and password. After the system validates the user details, it brings up information about the unit. The system provides a range of functions for entering jobs, reviewing the progress of jobs and providing feedback on a completed job. At this point, the unit manager will choose the function to enter a new job. For each of the common faults, the system will gather information about the symptoms of the problem, the effect of the fault on business activities, location of the fault and surrounding conditions. The system will get the input from the user through question/answer and multiple choice style interaction.

2. Based on the information, the system will identify the cause of the fault and appropriate contractor for the job. It will also determine a response time, i.e., 2 hours for urgent jobs, 24 hours or even longer for less urgent ones. The unit manager can override the system's suggestion on the understanding that quick responses usually cost more. The system then sends a message to the identified contractor.

3. If the system has difficulty in either understanding or interacting with the unit manager, the problem will be escalated and brought to the attention of a human helpdesk operator.

4. The operator will telephone the unit manager and interact in the conventional manner.

5. Through human intervention, it is hoped that the fault is accurately diagnosed. The operator enters the job details into the system and the new knowledge is captured for use in future cases.
6. When a contractor logs onto the server, all the jobs allocated to that particular contractor will be listed. Information about each job would include the location of the property, parking, opening hours, any health and safety information, nature and details of the fault. Based on the information, a suitable operative/engineer can be sent to the job.

After on-site evaluation of the job, if the estimated cost is within the agreed budget limit, the work will be carried out immediately. If the cost is higher than the limit, the contractor will submit an on-line quote through the system.

Once logged on to the server, the FM agent will be alerted instantly about the job waiting for authorisation. Detailed information about the job and quotation can then be accessed on the server. The agent can also access previous cases in making the authorisation decision. When approval is granted online, the work will go ahead, otherwise it will be cancelled. At this point, the Contractor will be alerted to the decision made by the FM Agent.

When the job is completed, the unit manager will provide some feedback on the quality of the work and workmanship of the operative. This information is used by the FM agent to assess the performance of the contractor.

In the meantime, the contractor will submit an invoice for the completed job. If the agent believes that the job has been completed satisfactorily, payment will be approved and made on-line.

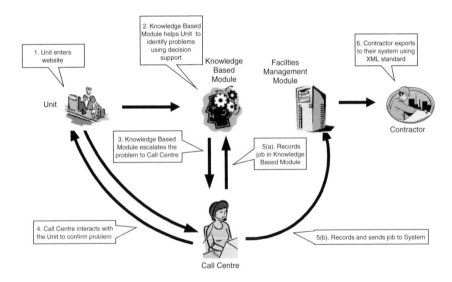

Figure 7.4 Maintenance job process using knowledge based on-line helpdesk system

7.4 FUTURE TRENDS

7.4.1 Flexible Working

Many changes in technology and management techniques will have an impact on the role of the facilities manager over the next few years. This role will increase in importance as the flexibility of working practices and the need for change increases.

Preliminary surveys show that the majority of workers in the UK are based in offices, either headquarters or satellites. Some employees spend their time out of the office either visiting clients or on assignments. The availability of mobile communications and Wide Area Networks has created an environment that allows organisations to be more flexible in their approach and increase the mobility of

---

Case study
Managing Property Portfolio Using CAD and Databases

Stoke Mandeville Hospital in Buckinghamshire has a building stock of the order of 80,000 square feet in 86 separate blocks. In 1985 when there was a mandatory requirement to update the fire alarm drawings for the hospital, it was found that information was either non-existent or inaccurate with much of it being spread out in different locations.

In 1993, the hospital was approached by Kinetic Technology who were promoting the 'Genesis' system, which comprised a CAD/Database package with a Paradox database. The package had a ready made link between CAD and a database. The price is also compared favourably to the mainstream competing products. The first workstation was acquired in 1994 by the hospital. The system has now grown to six computers. The strategy for the future is to continue to consolidate the system to improve its accuracy and reliability and get everything fully up to date.

Benefits achieved:

- Faster access to information. Modifications of drawings for development projects can be undertaken at the computer workstation with the client, providing considerable savings on project elapsed time.
- Building information is available at the touch of a button, rather than from paper drawings located at different parts of the site.
- Project budgets can be produced more quickly leading to more efficient management of the building stock and better planning of maintenance projects.
- Equipment owners can be shown the location of their equipment, reducing arguments about who owns various items and improving space utilisation.

Source: IT Construction Best Practice

their employees. However, the percentage of time spent in mobile or temporary office space is still small, but increasing.

It is safe to predict that those who select home working will not spend all their working hours there. Work will be divided between a fixed location such as a head office, for corporate meetings, and smaller satellite or managed offices where other resources may be used: powerful printers, video conferencing and smaller meetings, with some work being carried out at home.

The managed office (or tele-working village) may take on a more critical role. It has the attraction of reducing travelling time and providing shared resources, thus reducing capital outlay. The social aspect cannot be ignored since contact with others is important. Few have the resources for a completely equipped office at home.

7.4.2 The Future of Facilities Management

The implications of changes in the provision of buildings are significant. Firstly, sufficient space must be provided to accommodate all the services prospective occupants may require. Guidelines such as the CIBSE Guide to IT in Buildings detail the requirements for distribution of services in terms of space and power requirements.

The role of the facilities manager will be just as vital in 2015 as it is today, but it will be less related to a fixed work environment. In fact, in the next decade, facilities management will be a much more important function, encompassing many more activities. It could become more like a hotel service provision whereby workers book in when they need the facilities in the building. The concept of the virtual office is defined by its operation rather than its location. The virtual office can be accessed from anywhere in the country, mail can be sent to colleagues, documents viewed and printed, faxes sent and received and information exchanged.

Facilities management should be viewed in the future as an enabling function which creates and maintains an environment for an organisation to work efficiently and cost effectively.

7.5 INTELLIGENT BUILDINGS

This chapter has shown how important it now is for buildings to respond to their users' needs. Facilities management addresses the needs of the employer to make best use of the building and to control security and energy consumption. The occupant will be better served in future by more individually responsive systems.

Bill Mitchell (1995), in his book *City of Bits*, indicates that buildings are becoming more like computers. Control systems linking all parts of a building relay information between occupants and central plant and security. Environmental conditions can be monitored locally and corrected as necessary, ideally to suit the needs of each occupant. If a building is overheating in summer, blinds can be automatically lowered to reduce the load on cooling plant. As energy efficiency and environmental concerns become more important, these control systems will manage natural forms of lighting, and cooling techniques such as stack effect ventilation.

Already there are buildings with lights which switch on when movement or outside darkness are detected. Movement sensors can also be used for security purposes. All these local sensors and those for fire alarms can be linked to a single cable network reporting to a central Building Management System. The intelligent building of the future will know everything that goes on within it and will make automatic responses where necessary.

In the buildings themselves, processing power and data collection can be built into the fabric and services. Building management systems monitor comfort and pollution levels and adjust heating and ventilation accordingly. Lights can be switched depending on whether there is any movement from occupants, although this can prove difficult if blackout is required for showing slides.

Wireless networks, such as Bluetooth, allow communication between different machines not only in offices, but also in homes. A telephone call from the office can be used to switch heating systems, cookers and video recorders before the owner arrives home.

Intelligent systems are being built into a variety of things we use: car engines are controlled by them, aero planes are flown by them and buildings now have a growing number of control systems. These include energy management systems, movement and access control and security systems. They are often integrated with the data networks and can be used to find out who is in a building, what their environmental requirements are and how to locate them. To provide conditions to suit individuals requires very local control and distribution systems, but the advantages for each person's comfort may justify the expense of this.

The benefits for an organisation of having an intelligent building will lie in greater control of its use via a single building management system, saving of resources through providing resources only where they are needed and allowing staff to relocate more easily by moving their communications with them. Too much intelligence can alienate people, who should retain some privacy and control over their own environment. Video surveillance would be unacceptable except in very secure workplaces, but it is proving to be very useful in controlling crime in public places.

7.6 ON-LINE RESOURCES

Organisations

http://www.bifm.org.uk
The British Institute of Facilities Management (BIFM), established in 1993, is the UK's lead institute representing the interests of those who practice Facilities Management and those who work in organisations supplying Facilities Management related goods or services.

http://www.eurofm.org/site/
European Facility Management (EuroFm) is a Network of more than 50 organisations, all focussed on Facility Management. They are based in more than

15 European countries and represent professional (national) associations, education and research institutes.

http://www.rics.org/fm/
The RICS FM faculty is part of the RICS. It supports RICS members who are involved in facilities management (FM) and business support.

http://www.fmassociation.org
The Facilities Management Association (FMA) is a leading representative body for Facilities Management employers.

http://www.cfm.salford.ac.uk
The Centre for Facilities Management (CFM) is a network of blue-chip companies working together to further the understanding of FM and support applications in practice, research and education.

FM Publications

http://www.fmpages.net
FM pages is an Internet portal for FM professionals to find information on FM products and services.

http://www.fmuk-online.co.uk
Facilities Management UK is a glossy, high quality, high profile, business to business magazine published on a bi-monthly basis. Each issue responds to the facilities management market with in-depth intelligent articles covering important and relevant industry issues.

http://www.i-fm.net
iFM is an on-line portal for FM professionals. It maintains a comprehensive collection of up-to-date FM related information on news, events, technology and product development, and so on.

http://www.fmonline.co.uk
FM, another on-line magazine, is an FM portal.

CAFM Software Providers

http://www.archibus.com
ArchiBus/FM is a complete, integrated suite of applications that addresses all aspects of facilities and infrastructure management. The system is fully integrated with AutoCAD, ensuring that changes made to drawings are simultaneously reflected in the ARCHIBUS/FM database. It offers a variety of product options to accommodate different needs – from single users within a department to worldwide access over the Internet.

http://www.cafmexplorer.com
CAFM Explorer is an integrated FM system, which supports helpdesk, asset tracking, space management, property management, cost control and room booking.

http://www.serviceworks.co.uk/qfm/introduce.html
QFM, from Service Works, is a fully integrated facilities and estates management software package. It is a tool for the efficient management of a Facilities or Estates Department and is quick and easy to implement.

http://www.c2fs.com
FaciltyOne is an Internet based, web-enabled FM system. It provides its users quick and secure access to their facility systems documentation. It uses smart print technology to give maintenance and engineering staff the ability to review and analyse any facility blueprint with all relevant system data attached.

http://www.fdsltd.co.uk
Planet FM is a complete Facility and Estates Management solution that will help clients to optimise the utilisation of space, maintain assets and building fabric, and provide a responsive service to their occupiers.

SUMMARY

Facilities management is concerned with getting the most from buildings by adapting them to changing organisations. IT solutions link CAD and databases to record locations, services and assets, and allow these to be changed and reported on. IT is also important for control systems that are providing more responsive buildings and can report local environmental conditions, or fire and security problems. These can also be linked to business functions and be controlled remotely. The intelligent building of the future will respond automatically to all these needs and maximise the value of a building to its occupants.

DISCUSSION QUESTIONS

1. Before facilities management became a single function, who carried out its roles?
2. Consider the management systems used in your workplace. How is IT applied? What improvements could be made in this area?
3. What are the advantages of being able to operate buildings and equipment remotely?
4. Why is the database considered the most important part of FM software?
5. How could greater intelligence in control systems reduce building running costs?

CHAPTER 8

Integration

LEARNING OBJECTIVES

1. Understand why integration is important in construction.
2. Appreciate the possibilities of combining digital data.
3. Consider different method of integration.
4. Understand the need for product and process models.
5. Recognise the need for standards, both de facto and formal.
6. Anticipate the advantages of well integrated construction data.

INTRODUCTION

The process of designing, constructing and managing buildings is a complex one, involving different teams of specialists for each project. The data needed for each project is built up from the client's requirements, which are then given form during the design process, and the materials and work needed is priced and ordered during construction planning. Some of this data is needed to operate the building, and it has long been an objective to provide a means of transferring the information between each process stage and, as appropriate, right through a building's lifetime.

Integration is the process that would allow computer-based information to be passed between different types of software, running on different types of computer, and this should also be suitable for storage for periods of at least 25 years in an accessible medium. Now that all media can be in digital form it is possible to combine them as Multimedia and to use the best medium for each type of data. Problems of keeping data accessible over long periods have not yet been solved although storage media, such as optical discs, are robust enough, the means of reading them changes frequently. Now there is an awareness of this problem there should be services available in future to read data from any previous period.

There are different approaches to integration. The most obvious is for all software to obtain its input data from a shared database – in construction this would be a model of the building. However the fragmented nature of construction means that, unless a single point of responsibility is created, no one organisation has all the data, or is given the responsibility to build the model. Process changes in construction could provide solutions to this problem through partnering, Build Own and Operate and Supply Chain Management.

The alternative technique, which has been more successful to date, is that of gradual linking of individual software systems. In widely used packages, such as Microsoft Office, the facility for Object Linking and Embedding (OLE) has enabled data to be moved easily between the different modules. With design and analysis software for building, this has not proved so easy but translators exist between common types of software, and developments in Interoperability could offer a more complete solution.

Standards are essential to any exchange of data or interoperability of software. Hardware incompatibility has been tackled through the Open Systems Interconnection 7-layer model ensuring that different computers can communicate at all the levels necessary. Other formal standards include those for product models and CAD layers, as well as all the many building standards. However these formal standards can take a long time to develop, 7 years is typical and this is rather longer than a generation in IT systems. De facto standards, based on the products of leading manufacturers, have therefore been more successful in computing, but they can be exploited by their owners and rarely pass into the public domain.

It is possible to foresee a rosy future with the techniques available for integration being applied, and true interoperability of building data being achieved around an object-based model. However there are always reasons why these ideals may not be attained. Commercial differentiation of products is one. Each software supplier has a vested interest in not conforming to standards, which often represent the lowest common denominator. Most want to offer the standard but improve on it in ways that may limit interoperability. There is also the problem of who pays to build the model, and this will require greater commitment by the building client and a clear demonstration of the benefits that would result over the lifetime of a building. An integrated building process is necessary for IT integration techniques to deliver their full value in construction.

8.1 THE NEED FOR INTEGRATION

When mediaeval cathedrals were built there would have been a monk who decided upon the dimensions and decorative scheme, and a master mason who supervised the work which used traditional craft techniques. Communications were simple and the amount of data minimal even if the process could take hundreds of years. Since then construction has been getting more complex and the number of specialists involved greater. This required more formal communication, and conventions based on drawings, specifications and contracts became accepted and widely understood. These same types of information are still needed but can be 3D and in digital form, and capable of being combined or interlinked. This requires a new set of conventions that have not been fully developed yet.

One approach that fills some of the gaps, is represented by Coordinated Project Information, a project that resulted in the Common Arrangement classification system in the UK in the 1980s. This applies common codes to drawings, specifications and bill of quantities and allows cross-referencing. For graphical data the most common means of exchange between CAD systems is the external format of AutoCAD, DXF, but this is only effective for 2D data. A more radical solution is now required which recognises that buildings are 3 dimensional and that object-orientation techniques in software can be used to attach all the relevant data and procedures to families of objects. These objects, which may represent spaces, processes or components, can inherit data from others and the process of building up complex and repetitive models thus becomes more efficient.

Business Process Reengineering uses new technologies like IT to re-examine how companies could improve their productivity. In manufacturing industry, where the whole process may be controlled by one company and, even if there are external suppliers, it is possible to implement wholesale change. In construction the process is in the hands of several different companies. BPR can be applied to the operations of each of these but it is much harder to apply to the whole process. The Egan report looked at the process as a whole and suggested ways of encouraging greater involvement by clients and greater integration of project teams. Some of his recommendations are already being tried in the UK with partnering adopted by some large clients, and the government encouraging public/private partnerships. If these methods prove effective, the opportunity for integration of the building process will be greatly enhanced.

8.2 METHODS OF INTEGRATION

The ultimate goal of integration in an industry like construction is to have all the data needed by any of the participants interlinked. The problem is how to achieve this. If the industry was less fragmented and government or clients were prepared to invest in the necessary preparatory or standardising work, it might just be possible to achieve the ideal in 'one giant leap for mankind'. It is more realistic to try to achieve this more gradually – to integrate individual applications commonly used together step by step. Diagrams showing alternative means of integration are shown in Figures 8.1 and 8.2. For communications the centralised model in is more effective and is used in Project Extranets to make the same documents accessible to the whole team.

The question of which applications should be linked first depends upon how much data they share and whether they are used within the same type of organisation. Project management and construction management are normally the responsibility of the main contractor and share data on the tasks to be carried out. Design and analysis are typically carried out by engineers, while architects configure a building and specify its construction. Cost analyses are closely linked to tendering by the contractor, but also to the design and specification processes when carried out by quantity surveyors pre-contract. The stages will vary according to the type of contract and could eventually achieve full integration around a common set of data.

Integration has grown around successful software products. Originally professional offices used time sheet analysis/job costing software and contractors used accounting systems to manage their businesses. The successful ones, like SAP, grew and became Enterprise Resource Planning systems with many modules to suit any business need. In a similar way CAD systems like AutoCAD and Microstation have added modules for drawing management and data sharing. Each of these would like to become the main integration tool in businesses for which their software is central. However larger firms in construction often have several of these systems and, to control the integration and not become locked into one supplier, they may prefer to develop the integrating software themselves.

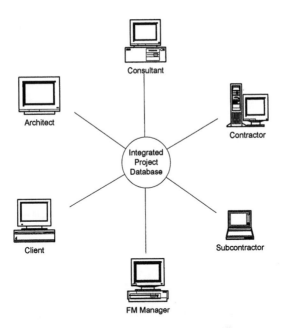

Figure 8.1 Total integration around a single project database

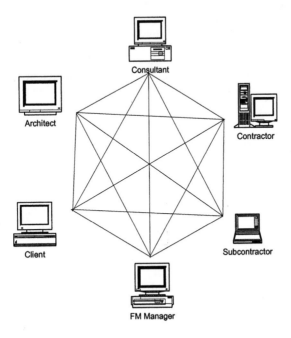

Figure 8.2 Communications between individual applications that can be linked gradually

8.3 PRODUCT AND PROCESS MODELS

8.3.1 Modelling

A model is defined as 'a representation of something' but, in computing, it is 'a simulation that describes how a system behaves so that a computer program can control the system or can explore the effects of change' (Eastman, 1999). The construction industry is used to physical models for visualising an unbuilt structure, but computer models can also be used to analyse performance and other characteristics. To operate on any conceptual design in a computer it must be modelled. This involves describing it using a language interpretable by computers, and specialist data modelling languages, such as EXPRESS, have been designed for this purpose.

Much early work in CAD concentrated on product models – representing the final artefact. Such models exist within most CAD packages but are mainly used to generate drawings and 3D visualisations. A model need not, but is generally understood to, contain 3D data, and may also have other, non-geometrical attributes. As techniques for representing the final built artefact in computers have become available, there has been growing concern with the processes for designing, making and managing artefacts, particularly buildings. Design and construction processes are very complex, involving many different specialists, and customers' needs cannot be satisfied without considering both the product and the process. Conventions are needed, both to ensure that models can be understood by different types of computer system, and so that they can be accessed and used by all members of the building team. Many of these conventions, which may be formal or de facto standards, have been defined, ranging from Open Systems Integration to those specific to construction, such as the Industry Foundation Classes. The development of a comprehensive building model, containing all the characteristics needed for presentation, analysis, documentation and life cycle management, has been a goal since the 1970s when integration of applications was first identified as important to allow computers to do more than automate existing design techniques.

8.3.2 Product Models

The most familiar computer models are those used for visualisation and contain only geometric information, using graphical techniques for rendering, photo reality, animations and interactive VR presentations. These help to communicate the designer's intentions to others but contribute little to the design process or analysis of performance. At one time it was hoped that building design could be automated by inputting the customer's requirements in the form of space requirements, and cost and performance criteria. Design requires many more factors, some of them cultural and hard to define numerically, to be taken into account and, so far, the human mind has proved to be superior. There are, however, uses of the bubble diagram techniques developed for automated layouts, such as reorganising office space between departments and floors of a building, known as blocking and stacking.

Data modelling languages have been developed over many years and, in 1964, SIMULA introduced the concept of objects. An object represents any thing and encapsulates rules for creating, manipulating and deleting it. It provides more security by requiring outside operators to call on these facilities to access it. An object also has inheritance in that it can share attributes with other objects. The significance of objects for building models is that they can represent whole buildings, spaces within them or the components of which they are made. They can contain all the attributes necessary for analysis, and their relationships with other objects. A door, for example, can inherit the characteristics of all other doors with variations in size or reference number. It can encapsulate its relationship with the wall in which it is placed so that the wall object knows there must be a hole and a lintel to take the weight of the wall above. Many CAD systems are now implemented in object-oriented languages such as C++ or Java.

All this technical capability to model buildings depends upon designers to supply the data and, where responsibility for design and construction is spread amongst several companies, there is a question of who benefits from better models. The client is the ultimate beneficiary from having more precise analysis and fewer errors through incompatible data or poor instructions to the contractor. The full benefit of modelling depends upon partnering with the client, or at least a single point of responsibility as in Design and Build contracts. Efficient modelling techniques and generally agreed conventions should make the modelling process more efficient and widely known, and this could help justify each of the project team's contribution to building the comprehensive product model.

8.3.3 Process Models

Modelling all the tasks, resources and stages involved in creating buildings started with workflow, bar charts and critical path techniques for controlling the construction process over time. A life-cycle approach can now be taken with the briefing and design processes also modelled, as well as facilities management and demolition. Much of the contractor's planning is concerned with 4D modelling, representing the construction process over time. Large housing developments have always planned use of labour so that the various trades move from house to house in sequence. Large inner city developments have to plan deliveries and storage of materials with limited space available.

When the main software applications for document production and performance analysis had been developed, it became apparent that further evolution into modelling could not be achieved without looking critically at industry processes. The Latham and Egan reports in the UK looked at relationships between those involved in building and set targets for productivity improvements. Many of the means of achieving these have now been tried out by the leading firms and are documented by the Construction Best Practice Program. IT and modelling systems have been the catalyst for some of these changes, and the combination of better models and process changes should eventually help increase productivity.

8.4 STANDARDS

8.4.1 Data Model Standards

Historically, the initial requirement for a standardised data model came from the need for different versions of CAD applications to exchange their graphic files. IGES (the Initial Graphics Exchange Specification) of the United States was developed for this purpose. However, graphical and geometrical data is only part of the information required in a building project. IGES was not able to support the exchange of other types of data such as construction, thermal, lighting, etc. Therefore a new project, PDES (Product Data Exchange Specification), was proposed in the US in the early 1980s to overcome these limitations. In the same period, similar efforts were made in other countries, for example, the SET (Standard d'Echange et de Transfert) in France and the VDAFS (Verband der Deutschen Automobilindustrie Flaechen Schnittstelle) in Germany. In 1983, all these initiatives were co-ordinated into a major international program under the umbrella of the International Standards Organisation, the Standard for Exchange of Product data (STEP). STEP defines not only standard data models to facilitate information exchange but also a standard methodology for data modelling and data exchange. The STEP standard is still evolving. It has a very ambitious aim of defining standard data models for all manufacturing sectors, such as the aerospace, automotive and AEC industries.

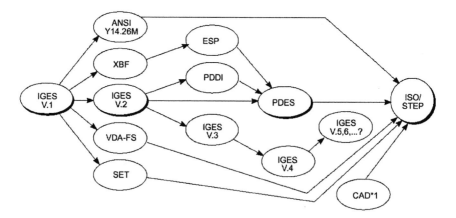

Figure 8.3 The evolution of data exchange standard (after Warthen, 1989)

The lack of integration between different commercial AEC applications has become a barrier to the wide use of these systems. To address this issue twelve companies involved in the AEC and Facilities Management (FM) Industry, many of whom are software developers, started the International Alliance for Interoperability (IAI) in 1995 (IAI, 1999). The aim of the IAI is to achieve software interoperability in the AEC/FM industry. The mission is to define, promote and publish a specification for sharing data throughout the project life cycle, globally, across disciplines and across technical applications.

In essence, IAI is another data modelling initiative which seeks to specify how the 'things' that could occur in a constructed facility (including real things such as doors, walls, fans, etc. and abstract concepts such as space, organisation, process etc.) should be represented electronically. These specifications represent a data structure supporting an electronic project model useful in sharing data across applications. The IAI adopted the object-oriented paradigm. The specification of each type of real world object, such as doors, walls, windows, is called a 'class'. The IAI model is a collection of classes, which are called 'Industry Foundation Classes' (IFCs).

8.4.2 Data Modelling Standards

Data modelling has always been an integral task in the software engineering process. There are several established modelling techniques, both graphical and textual, such as entity-relationship method, data flow diagram, NIAM, IDEF-1x, etc. With the increasing popularity of the object-oriented paradigm and the growing influence of the STEP initiative, EXPRESS has emerged as the de facto standard data modelling language (ISO, 1994).

The first task of data modelling is to define the scope of the model. The building system is a highly complex one. At present it is impossible for any single data model to cover the whole building life cycle and all processes associated with it. Each model needs to define the scope it intends to cover. The EXPRESS data modelling language is in compliance with the object-oriented paradigm which, in recent years, has had a major impact on data modelling activities. Major principles applied in the modelling process include abstraction, generalisation, aggregation, association and decomposition.

8.4.3 Data Mapping Standards

The standard data model is intended to aggregate the data requirements of all the disciplines in the life cycle of a building. In practice each application only uses a subset of the model and has a partial view of it. It is therefore necessary to define some standard mechanisms to map these data views of different software application. So far a number of data mapping standards have been developed including EXPRESS-M, EXPRESS-V and View Mapping Language (VML).

EXPRESS-M was developed initially to solve the problem of Application Protocol (AP) interoperability in the STEP standard. It is a schema mapping language used to describe how entity instances should be mapped between schemas in order to facilitate the transfer of data between models described by those schemas. EXPRESS-M is designed to allow the mapping of disparate EXPRESS defined data between applications with STEP Part 22 compliant interfaces. EXPRESS-M mappings are unidirectional and map a whole model at a time, not partial updates of models.

EXPRESS-V was developed to enable the user to define simpler views of a complex and integrated data model in EXPRESS format. The language begins with the definition of two EXPRESS schemas. The first is the source schema and the second is a target schema that defines the entities to be included in a view of the source schema. An EXPRESS-V mapping schema is then defined to specify how entities in the source schema are to be mapped to entities in the target schema, to create an instantiation of the view from an instantiation of the source schema. An application system that uses the entity instances defined by target schema is likely to make changes to those entity instances. Therefore, EXPRESS-V defines not only a mapping from the original information model to the view, but also a mapping from the view back to the original information model so that changes can be propagated back.

VML is a bi-directional, high level and declarative language for the specification of mappings between two arbitrary schemas or versions of schemas. A VML mapping system can be configured to perform in an interactive environment as well as handling full model conversions. It allows the simultaneous connection of multiple applications or databases and accepts incremental modifications to a model, which can be propagated through to all connected applications. Such a system can therefore verify the global consistency of models in an integrated system by tracking that modifications have been propagated and which changes are outstanding for an application.

8.4.4 Data Exchange Standards

In a computer integrated construction environment, data exchange is commonly understood to be the exchange of neutral format data files between computer systems. A sending system translates data from its internal format and encodes it into an established neutral format. This file is then transferred to the receiving system where the data is translated into the internal format of the receiving system. DXF is a well-known file format for exchange of graphic files between CAD systems. The STEP physical file format has emerged as the neutral format for exchanging full product data.

A STEP file is, in essence, a text file that contains both values of data and implicit knowledge for interpreting the data. The systems involved in the data exchange have access to a conceptual data model that serves as an explicit standardised data specification for the STEP data. It is this model that provides a documented explanation of the context (scope) and meaning (relationships) of the data to be exchanged. It is used, along with an encoding algorithm, to produce STEP physical files that contain both the data and its associated context, thus enabling effective and flexible communication between computing systems.

The Extensible Markup Language (XML) is a meta-markup language that provides a format for describing structured document which contains content and knowledge of what role the content plays (Bryan, 1997). Structured documents can be spreadsheets, address books, configuration parameters, financial transactions, technical drawings, etc. This facilitates more precise declarations of content and more meaningful search results across multiple platforms. In addition, XML will

```
ISO-10303-21;
HEADER;
FILE_DESCRIPTION($);
FILE_NAME($,'Thu Nov 26 15:28:08 1992',('PdV'),('TUD'),'IDM_Instool
v.0.1','IDM_DEMO',$);
FILE_SCHEMA('IDM');
ENDSEC;
DATA;
#1 = CATALOG('DTP1(353)','DTP1 Construction Library');
#2 = CONDUCTIVITY('DTP1(800)',0.57,$,'Thermal conductivity','Real');
#3 = EMISSIVITY('DTP1(801)',0.95,$,'Thermal emissivity','Real',( ));
#4 = REFLECTANCE('DTP1(802)',0.65,$,'Light reflectance','Real',$);
#5 = SPECIFIC_HEAT('DTP1(803)',840,$,'Specific heat capacity','Real');
#6 = MASS_DENSITY('DTP1(804)',700,$,'otherwise known as density','Real');
#7 = MATERIAL('DTP1(503)Gypsum plasterboard on plaster
dabs',#2,#3,#4,#5,#6,$);
#8 = LAYER('DTP1(502)InternalWallFinishes',$,0,0,0.01,0,#7);

...

#324 = ENCLOSURE_ELEMENT('338',$,$,#323,$,(),$,());
#325 = ELEMENT_CONSTRUCTION($,0,$,$,0.3,#23,());
#326 = ENCLOSURE_ELEMENT('Encl_Elem_299',$,$,#325,$,(),$,());
ENDSEC;
END-ISO-10303-21;
```

Figure 8.4 An example STEP file

enable a new generation of Web-based data viewing and manipulation applications.

8.5 INTEGRATED PROJECT DATABASES

In parallel to the above mentioned data modelling activities, there are many research projects world wide which investigate the integration of AEC applications through an implemented project database. The aim of an Integrated Project Database (IPDB) is to provide consistent and reliable storage of the project information, and serve as a data exchange hub for different tasks during the construction process. The requirements for an integrated project database can be summarised as follows (Sun, 1997):

- *Persistent building model*: Building design is a long process during which information about the building is enriched gradually. The first requirement for the IPDB is to hold a persistent model for different states of the building.
- *Data exchange interface*: The second requirement for the IPDB is to support data exchange interfaces with third party construction software through the use of product data technology.
- *View Integration*: Each software package deals with a particular aspect of the construction problem, thus, it has a partial view of the building model. These partial views need to be merged into a single coherent building

model in the central data repository. This determines the requirement to maintain a consistent building model and to support the growth or population of this model through the integration of partial views.

- *Change management*: In a real world construction project, if the architect feels the need to make changes to the building layout, he or she would inform the HVAC engineer to halt the heating and cooling system work, since they will have to be re-designed as a result of the building geometry changing. In a computer based design support system, a degree of concurrent engineering management is expected. This determines the requirement of the IPDB to provide the underlying support for change management and conflict prevention.
- *Design versioning*: An essential feature of any design process is the requirement to explore alternative design solutions. Design versioning should allow each discipline to take a version of the design at any stage. These versions will develop independently and co-exist. They also need to be merged to produce one coherent version of the design.
- *Project history*: An IPDB should be able to record the project history trail and allow professionals to re-examine the decision making process and to roll back to an early consistent state of the project.

Amor and Anumba (1999) presented a survey of six major EU and UK research projects that involve the development of IPDBs. He revealed that there is a consensus on the conceptual approach and the object oriented database implementation of the IPDB amongst these projects. However, the scope of these existing IPDBs is still limited and a cradle-to-grave IPDB will not appear in the near future. The IPDB scalability and the interface with application software still pose issues for further research.

8.6 AN EXAMPLE INTEGRATED SYSTEM – GALLICON

GALLICON (Sun, 2000) is a research project jointly funded by the Department of Environment, Transport and Regions (DETR) and a consortium of partnering companies, Galliford, EC Harris and Welsh Water, in the UK, who are working together on many water treatment projects. At present the design, cost estimating and project planning tasks are carried out using computer software independently during the design appraisal. Due to data incompatibility, the information exchange between the tasks is still done using a paper medium. Given these tasks are often geographically distributed, the traditional paper based information exchange is often a cause for project delay and errors. Furthermore, due to the lack of integrated information management, the decision-making knowledge for each project is not captured effectively nor is it stored in an appropriate format to be reused in future projects. The main aim of the GALLICON project is to develop an integrated information framework which will improve communications and information exchange between the distributed processes of design, cost estimating and project planning.

8.6.1 The GALLICON System

Figure 8.5 shows the system configuration of the prototype developed during the GALLICON project. It adopted a distributed architecture that consists of a repository which is a shared project database, a System Administration application, interfaces to three third party software packages, and a VRML browser. The application software packages are used for design, cost estimating and project planning. The project information is created by the application software and stored in the project database, which can be accessed by all applications of the system. The VRML browser provides a graphical visualisation tool for the project data, which facilitates the communications between partners located at different places.

These applications operate in an integrated manner; changes in one of them are reflected by changes in the others. This means that, for example, by creating a layout design in the CAD application the cost estimating and project planning applications automatically generate cost and planning information related to that design. The purpose of the System Administration is to manage user access to the

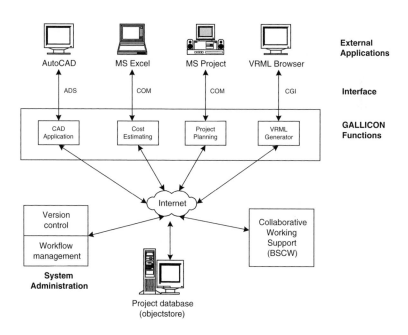

Figure 8.5 System architecture of GALLICON

project database and to control the operation sequence of the system based on workflow information. The Virtual Reality Modelling Language (VRML) browser displays design layout, cost, and planning information over the Internet in a virtual reality environment. The information is generated automatically using data from the project database. This facility is particular useful for the clients to enable them to play a more active role in this design appraisal process.

The integration of the CAD, cost estimating, project planning, and visualisation applications in GALLICON, is achieved through a central project database. Each one of the applications generates part of a specific object, such as a 'settlement tank' object. These data are stored in the central database in a common format that is rich enough to accommodate the needs of all the GALLICON applications.

The CAD application is the starting point from which new construction objects are created or existing ones are modified. Construction objects are created and drawn in the CAD application by calling purpose built commands and not by using native AutoCAD commands. For example, to draw a nitrifying filter in the wastewater treatment prototype the designer calls the command 'Draw Nitrifying Filter'. He or she then specifies a series of attributes, such as the filter's position, geometry, and composition, and the system draws a representation of the filter using basic AutoCAD objects, such as lines and circles. This representation is the footprint of the construction object. However, a 3D representation is generated in the project database even though it is not shown on the CAD drawing. Each object in CAD is an information rich object that has specific attributes and behaviour. It knows its role in the design and its relationship with other objects. For example, a tank object knows its possible connection points with a pipe, if it can be connected to one. To connect two tank objects with a pipe the designer only needs to select one of the possible connections points presented by the system. When a tank object is moved to a different position, the system will automatically readjust the connecting pipe so that the two tank objects remain connected.

Each time a new object is added in CAD or an existing object is modified, the cost estimating and project planning applications read the new properties and perform various calculations under the control of the user. By having access to detailed information about each design object and related information on external factors, such as the geology and topology of the project site, the cost estimating application calculates project costs by applying quantity takeoff rules. However, GALLICON does not aim to automatically generate a complete and absolutely accurate cost estimate, because to do so requires huge amount of input, some of which is simply not available at the early design stages. Instead, it seeks to assist the user in calculating some of the costs in order to perform what-if analysis where comparative costs are more important than absolute ones. The system also seeks to improve the information flow between the project team members by allowing information sharing through the project database and change notification functions. During the operation, the user has control over when and if a change should be sent to the project database. Once a change is made, the system will automatically inform other members of the change and request new calculations when necessary.

In a similar way, the project planning application does not seek to generate a full project schedule. To do so would require addition information, such as plant and labour resources and their availability. What GALLICON does generate, however, is a list of all the tasks associated with a project and the durations of these tasks based on the assumption of standard personnel and equipment resources. The unit costs of materials and labour are taken from a standard Price Database used by EC Harris. Interface with other unit cost databases is feasible. The project manager has the overall control to accept or change the generated tasks and to create any schedules.

8.6.2 Design Collaboration Using GALLICON

The GALLICON system provides support for collaboration by a multi-disciplinary project team located in different parts of the country; some may be on-site, others located in offices. All team members need to have access to a computer and the Internet. Computer based communication and collaboration tools are used to connect the project participants and allow them to interact with each other and share information. In the current prototype, a WWW based virtual workspace is implemented based on the Basic Support for Collaborative Work (BSCW) prototype (Bentley, 1997). The system enables remotely located team members to exchange textual messages, and share workflow information. It also co-ordinates the access to the integrated project database using the process model knowledge. A design collaboration scenario using GALLICON can be described as follows:

At the start of a new water treatment project, the project manager will set up the integrated project database on a central server and give all participants, designers, cost estimators, planners and clients, appropriate levels of permission for access. Then, the designer will start an outline design using the customised AutoCAD tool at the design office. The CAD tool also allows the specification of materials and components using data from a standard product library. Once the designer is satisfied, the information can be saved into the project database on the central server in the same office or in another location. The project manager is informed of the progress.

When the project manager logs on to the system in another location, through a web interface, he or she will get the message that a design option is proposed. The manager will in turn inform the cost estimator to carry out a cost analysis. The communication is done through the on-line message facilities. The cost estimator, who may be in yet another office, can access the design information from the central project database through the Internet. The cost estimator is able to see the exact status of the project visually using a VRML viewer. The cost estimating tool will generate some quantities automatically, such as the amount of excavation, and give instant access to standard material and labour unit cost information. However, it cannot and does not intend to replace the cost estimator altogether. The overall effect is to eliminate routine calculations so that the estimator can concentrate on those tasks that require knowledge and judgment. The designer and cost estimator can work collaboratively and interactively to modify the design and component specifications.

At the end of the design phase, the project planner is asked to create a schedule of work. Traditionally, manual calculations are needed to work out all the work items and the duration of each item. In GALLICON, the system does the calculations automatically using design data from the project database and a standard cost library. The calculation results are presented as a list of tasks in MS Project. The planner can generate a schedule of operation based on the information. Of course, the user can change the generated results using the full functionality of MS Project, for instance, changing the length of a task duration, adding new tasks or removing tasks.

Communication is the most important aspect of multidisciplinary collaboration. The GALLICON system allows the project team not only to share project data, but also to share the rationales of the decision-making. When design changes occur, the Internet messaging function and the BSCW support will inform the other team members:

What has been changed
What is the reason for the change
Who changed it and who agreed those changes
When has it been changed
What are the implications of these changes

The decision history trail is recorded and can be reviewed at a later stage. All communication is done through a web-based tool. The system allows viewing of a variety of documents. The GALLICON system provides a virtual reality graphic interface to the project database. All the team members, including the client, can view all aspects of the project including design, costs and project programmes, from their own desktops. This is particularly useful to involve the client early in the design process.

8.6.3 Benefits of the Integrated System

The industrial partners of the project were interested in measuring the benefits of the developed system in real projects. In fact it has been recognised that the lack of benefit measurement of IT innovation is one of the most common barriers against construction companies making IT investments. In the construction industry, where profit margins are usually tight, a common trap in measuring IT benefits is to focus purely on monetary effects, i.e., how much money the system will help to save. IT investment should be considered as a long term investment with multi-faceted benefits, such as improvement in client satisfaction and perception, gaining competitive advantage, more efficient business processes, etc. Baldwin (1999) presented a practical methodology for measuring the benefits of IT innovation. It proposes that any assessment should be conducted in three areas: efficiency, effectiveness and performance. Efficiency is financially measurable and is represented by money. Effectiveness is measurable but not in monetary terms and is represented by improved operation. Performance is not directly measurable in quantifiable terms but can be judged qualitatively by its influence on long-term business performance by increasing profit and market share.

The benefit analysis of the GALLICON system was conducted through benchmarking workshops involving a panel of experts. During the workshops, the researchers demonstrated the system for a sequence of tasks using a sample project. The purpose of the demonstration was to stimulate discussion by the panel members. The panel members were invited to comment on the potential benefits of the system using the current approach as a benchmark. The findings, summarised in the following, are qualitative rather than quantitative:

- At the highest level it was felt that the communications costs could be reduced significantly.
- In the partnering approach, face-to-face meetings were considered important. However, the project partners felt that, with the quality of information being presented by the system, these meetings could be reduced. In addition, if some team members were unable to attend the meeting, they could communicate more easily via the system.
- The cost estimates automatically generated by the system have an

---

Additional Information
A Vision for Future Integration

The client for a commercial complex of the future has commissioned similar ones before and his briefing information is well developed. He has also set up profit, and risk, sharing partnering agreements with consultants and contractors to work with him on a series of projects. An information manager is responsible for controlling the sharing of data on the project Extranet. The contractor is appointed on a fee basis at an early stage, as are many of the subcontractors and suppliers.

A meeting is held to agree responsibilities and formats for the project model. This is based on the Industry Foundation Classes and time is given for building a comprehensive 3D model before any work starts on site. This model is fully analysed and the performance of the design simulated. The construction sequence is tested using 4D techniques and detailed models of particularly complex areas. As many potential faults as possible are picked up during the design stage.

The design is explored with the client using VR so that the people who will work in the building are fully aware of its configuration and can suggest changes before it is too late. Construction proceeds with all those involved accessing the Extranet into which are built many of the standards to ensure integration. On completion the data needed for facility management is retained by the client, and the project team shares the rewards for completing a quality project on time and within budget. The building carries with it for its lifetime data on its materials and performance so that, when it is finally demolished, its materials can be recycled or disposed of safely.

acceptable accuracy. It does the calculations in minutes as opposed to days using manual methods. The accuracy can be improved through the manual intervention of the cost estimator.

- The panel of experts viewed the use of the Integrated Project Model approach, as 'a huge leap forward, there is nothing around like it'. The information blocks between the project team were widely recognised and it was felt the system addressed these issues.
- The use of the 'what if?' analysis approach was viewed by the panel as a possible driver for change in the way projects are constructed. The system allowed model experimentation rather than choices based on experience, which may be subjective.
- The more open use of information and the logging of rationale create greater opportunity for individuals to participate in the 'value engineering' activities. Additionally, the system records the rationale for these reasons, which may be reviewed in future projects.
- The use of the release and version control data informs all partners what changes have been made, keeps the most up-to-date version available and archives older versions. This system operates as a strong document control system for the project, ensuring all partners are working on the same data.
- The client felt there was a strong qualitative improvement in the information being presented. The partnering approach led to a reduction in information due to increased levels of trust. Giving the client access to multiple views of 'live' project information improves the client's perspective, whilst supporting the open book ethos of partnering.

8.7 ON-LINE RESOURCES

Standards

http://www.iso.ch
The International Organisation for Standardisation (ISO) defines ISO 10303 Standard for the Exchange of Product (STEP) model data.

http://www.iai-international.org
The International Alliance for Interoperability (IAI) is a worldwide consortium aiming to define the requirements for software interoperability in the AEC/FM industry. The deliverables of IAI are the specifications of the Industry Foundation Classes (IFC) – an object oriented software library for application development.

http://www.xml.com/
A site dedicated to the XML standard and its applications.

Applications

http://www.blis-project.org/
Building Lifecycle Interoperable Software (BLIS) is a group actively implementing IFCs into commercial software products, defining use cases and testing the interoperability between their products.

http://www.iai-na.org/aecxml/mission.php
aecXML is an XML-based language used to represent information in the Architecture, Engineering and Construction (AEC) industry.

http://www.xmlsoftware.com/
A site dedicated to software using the XML standards.

http://www.steptools.com/
STEP Tools Inc. is a developer of STEP standards-based software products for e-manufacturing.

http://www.eurostep.se/
Eurostep is a consulting and software company specialising in information management.

Research Projects

http://cic.vtt.fi/projects/index.html
Links to many EU research projects on computer integrated construction.

Portals

http://cic.vtt.fi/home/links.html
Comprehensive website links related to integrated design and construction maintained by VTT in Finland.

http://w78.civil.auc.dk/
CIB W78 – Information Technology in Construction.

SUMMARY

As with many developments in technology it is often easier to see what the final scenario might look like rather than to understand the steps necessary to get there. A vision of a project of the future, applying all the integration that should be possible, will serve to show why this is such an important goal.

This may seem like a pipe dream but the technologies needed to achieve it are largely in place already. What is missing is the human will to achieve it and the resources and coordination needed from an understanding client. Some of the process changes already being tried out are providing convincing evidence of the potential benefits of integration. Only when these are all applied to a suitable project will the full benefits of technology and process change become attainable.

DISCUSSION QUESTIONS

1. Discuss the relative merits of gradual integration of applications and the use of a common building model.
2. What are the characteristics of product and process models?
3. Describe some of the attributes of objects used in object-oriented systems.

CHAPTER 9

Future Developments

LEARNING OBJECTIVES

1. Gain knowledge of the trend of technology development.
2. Recognise the potential applications of the new technologies.
3. Appreciate the future vision of IT applications in construction.
4. Know the basic steps of planning and implementing a vision.

INTRODUCTION

Developments in information technology have focused on each of the elements of a computer system in turn. First it was the hardware that had to be made compact and reliable, then software needed to become easier to use, and then communications made it possible for what had been a specialised technology to become available to all. Even those who do not wish to use computers now find themselves doing so through their mobile phones or digital television sets.

Amazing achievements have already been made in miniaturisation and cost reduction. For example the Atlas II computer at the CAD Centre in Cambridge had 128K of memory and occupied a whole suite of offices in 1970. By 1995, the first PDA, the Apple Newton, offered 1 megabyte in a palm-sized device. Moore's Law stated that the number of transistors, and thus the power, of an integrated circuit would double every 18 months while the cost remained the same or even reduced (Turban, 2002). The Intel Merced chip has over 10 million transistors and operates at about 500 MIPS, millions of instructions per second.

The pace of technological advance is accelerating. More powerful computers and better connectivity will provide a strong 'technology push' for wider use of IT in construction. In this chapter, we review some of the latest technology developments and their potential applications. We will also discuss the industry's drive for performance improvement, or the 'business pull' for IT use.

9.1 TECHNOLOGY FORESIGHT

The UK government has an on-going Foresight programme, started in the 1990s. Its aim is to identify future opportunities in market and technologies in different sectors of the nation's economy. One of the panels of the Foresight programme is the Information Communication and Media (ICM) panel. In 2000, the ICM panel set up an IT, Electronics and Communications (ITEC) group, which was charged to identify the development trends in the ITEC sector over the next 15-20 years.

The ITEC group's study suggested that the pace of technological change in IT and electronics industry has been accelerating. In the next decade, we will witness significant advances in computing and communication technologies. The following discusses several areas where these advances are expected.

9.1.1 High Speed Network and Mobile Network

In recent years, the access to computing networks has been widening. Most businesses and many homes in the UK have joined the Internet community. However, the transmission capacity of the current network connection is still relatively slow. A modem dialup connection, with a maximum transmission rate of 56k bits/second, is only suitable for browsing non graphic-intensive web pages. Its advantage is that it works on the existing telephone line and is relatively cheap. As a result a modem is the most widely used method of connecting to the Internet, especially for home users. However, its transmission rate is too slow for transferring multimedia contents and large data files.

The Integrated Services Digital Network (ISDN) was the earliest alternative to a modem. The transmission rate of ISDN is about twice the modem speed. It uses dedicated lines. Therefore there is more guarantee of getting the maximum transmission speed. However, because of the additional lines and devices, ISDN is expensive to install and operate for the marginal performance improvement. With the recent emergence of Asymmetric Digital Subscriber Line (ADSL), ISDN has been gradually phased out.

ADSL delivers data over the existing copper telephone lines using advanced data compression. It can deliver content about 50 times faster than a dial-up modem. Typically, ADSL services provide speeds of greater than 2.2 Mbps downstream (from the network to the personal computer) and up to 1.1 Mbps upstream (from your computer to the network). The service does not require a separate phone line and is always on. This means a user can be on the phone and access the Internet at the same time on the same line. 'Asymmetric' refers to the different speeds of two-way communications. This makes ADSL most suited for home users who mainly download information from the Internet while uploading less information. ADSL is less suited for business users who require similar speeds for two-way communications. Symmetric Digital Subscriber Line (SDSL) has the same transmission speeds in upstream and downstream directions over a single copper-wire pair. It is slightly slower than ADSL, but has equal speed in each direction. Therefore, SDSL is better equipped for videoconferencing and other business applications that involve extensive interactions.

Cable modem technology uses the existing infrastructure of coaxial wire used to deliver cable television. However, cable television systems were originally designed to transfer signals one-way, that is, to deliver channels into the home. To deliver data services, the cable companies have begun to upgrade their network with the digital switches necessary to provide bi-directional communications. The cable company's advantage over the phone companies is that the cable network is largely fibre optic with coaxial cables running into subscriber homes. Coaxial cable is much thicker and more amenable to very high bandwidth uses than the thin copper cables used in phone systems. The current generation of cable modems provides downstream bandwidth ranging from 10-30 Mbps, roughly 350 times faster than today's modems, and modems providing over 100 Mbps have been tested. Cable modems are also asymmetric in that they receive data at a higher speed than they send. Current cable modems receive data at about 10 Mbps and send at about 800 kbps. The theoretical maximums for coaxial cable are much greater than for copper wire and there is some expectation that gigabit (1000 Mbps) cable modems will become available in the future.

Table 9.1 is a comparison of different network connections, their transmission speeds and some representative applications.

Table 9.1 Network capacity comparisons

Type of connection	Transmission rate (bits/second)	Typical applications
Dialup modem	Up to 56,000	Browse WWW pages with text and static graphics
ISDN	128,000	Low resolution point-to-point video conference
ADSL	500,000- 2,500,000	HiFi quality sound and VHS quality video on demand
Cable modem	12,000,000- 1,000,000,000	Full screen DVD quality video

The biggest limitation of fixed line networks is that all connection points are in permanent positions. The construction industry has many mobile workers travelling between offices and building sites. They would like to have access to the network while on the move. Mobile or wireless network technology can provide the solution.

Wireless networks use radio waves as a transmission medium. Transmission signals are relayed by either terrestrial transmitters or satellites. One example is the 3rd Generation (3G) mobile technology. 3G uses Universal Mobile Telecommunications Service (UMTS) technology and allows mobile users to access integrated voice and data communications services. At present the bandwidth of this type of network is still limited. Contents need to be prepared as a 'light' version of the normal HTML pages compliant with the Wireless Access Protocol (WAP). In the longer term, i.e., 2008 onwards, there are plans to develop Mobile Broadband Systems (MBS) with data rates above 2Mbps.

Another area of networking that is expected to grow rapidly over the coming years is personal area networking. With the development of new technology such as wearable computers, enhanced personal digital assistants (PDAs) and data centric mobile telephony, there will be a need to network these products together. Wireless technology such as Bluetooth is predicted to play a key role in personal area networking. Bluetooth enabled devices to be able to communicate with each other over short ranges, such as in a home or an office.

9.1.2 Pervasive or Ubiquitous Computing

Pervasive or ubiquitous computing is a term used to describe the trend that computers are moving away from the desktop into the surrounding environment. It is brought about by a convergence of the miniaturisation of advanced electronics, wireless technologies and the Internet. Pervasive computing devices are not personal computers as we know them, but very tiny, even invisible, devices, either mobile or embedded in almost any type of object imaginable, such as cars, television sets, refrigerators, clothing, mobile phones, and other electronic devices. These devices will have information about their identity, current location and key functions. They can communicate with each other when necessary without the need for explicit instructions from the users. According to Dan Russell, director of the User Sciences and Experience Group at IBM's Almaden Research Center, by 2010 computing will have become so naturalised within the environment that people will not even realise that they are using computers.

The main goal of ubiquitous computing is to create systems that are pervasively and unobtrusively embedded in the environment, completely connected, intuitive, effortlessly portable, and constantly available. This technology will have a direct application potential in construction and buildings in the forms of wearable computers, smart homes and smart buildings.

At present, this technology is still at the research stage. Many world leading research centres have been conducting research since the 1980s. These include the IBM's Planet Blue project, the project Aura at Carnegie Mellon University, the project Oxygen at the Massachusetts Institute of Technology and the on-going work at Xerox Palo Alto Research Centre. The research focuses on solving a number of problems before pervasive and ubiquitous computing can become reality, such as application-specific integrated circuitry (ASIC), speech recognition, gesture recognition, system on a chip (SoC); perceptive interfaces; smart matter; flexible transistors; reconfigurable processors, and microelectromechanical systems.

9.1.3 Intelligent Speech Recognition

In the last twenty years, computers have improved tremendously. However, the user interface technology, especially user input devices, has changed little. Keyboard and mice are still the primary method to enter information into computers. Voice recognition is one of the new ways of interfacing between humans and computers. At present, many desktop voice recognition systems can convert speech to text fairly accurately, some claiming an accuracy rate of 95%. However, it does require lengthy training for the system to recognise a particular speaker. More discouraging, it usually also requires reverse training. In other words, the users have to change their natural speech styles to suit the system.

In the future, speech recognition techniques, in addition to their use for dictation purposes, are expected to be widely used as methods of verifying peoples' identities in telephone banking and shopping services, information retrieval services, remote access to computers, credit-card calls, etc. To achieve this, service providers will store customers' speech signatures in the same way as current hand written signatures. Then, a customers' voice instruction will be quickly compared with the stored signature to verify identity.

The same technology will be applied to the control of buildings. Occupants will be able to do things using voice commands to the ubiquitous computing devices embedded in the building fabrics, for example 'open door', 'switch on light', and so on. This will be particularly beneficial for people with disabilities.

9.1.4 Telepresence

People can communicate instantaneously with each other over long distances using telephones and videoconferences. However, the existing technologies cannot create a sense of 'being there' or 'face to face'. Current systems typically deliver at best a 'through the window' experience. The future will bring progress towards 'through the screen' telepresence, where the user will feel increasingly as if they were passing through the screen into the remote location, rather than merely observing it through a window. There are two aspects of telepresence: co-presence and remote-presence. The co-presence aspect enables people to meet in virtual space as if they were in the same physical space. The remote-presence aspect allows people to carry out work from a distance, for example a surgeon could carry out an operation remotely.

One approach to achieving telepresence (co-presence) uses large screen displays, perhaps with multiple projectors, and a mix of techniques from the traditionally distinct disciplines of computer graphics and computer vision. For example, it is possible to create the illusion of a room or conference table that extends seamlessly through the screen, linking two or even three physical locations. Another method uses personal devices such as headsets. Increased display resolution, reduced lag in tracking, and less physically cumbersome designs, are making these a credible commercial solution.

One example of a telepresence system is provided by the EU VIRTUE project. The aim of the project was to develop an improved videoconferencing system so that a sense of co-presence could be achieved between the meeting participants. A semi-immersive videoconference station was developed with a large plasma display and four cameras mounted around the display (Figure 9.1). It enables three conferees (insert in Figure 9.1), one local and two remote, to participate in real-time meetings. Correct eye-contact was achieved by creating virtual cameras behind the display screen. The video streams from these virtual cameras were created in software from the video output from the four real cameras.

9.1.5 Computing Grid

The current generation Internet provides a successful platform for information dissemination and electronic commerce. However, as more users join the network, its bandwidth limitation becomes increasingly acute. Now, there are several initiatives worldwide, which seek to develop the next generation Internet using

Figure 9.1 The VIRTUE telepresence station (images courtesy of the BT's VIRTUE project)

computing grid technology. The new computing grid technology will allow seamless access and use of computing resources as well as information. The examples of this new development include the Internet2 initiative and the NASA Information Power Grid in the USA, the European Grid (E-Grid) in Europe and the e-science/e-grid project in the UK.

The e-science project started in 2000 with £120m funding from DTI, research councils and industry. Its initial aim is to set up a computing grid infrastructure covering the whole country and demonstrate its application through pilot projects. At present, in addition to the National Centre in Edinburgh, there are eight regional centres based at the Universities of Newcastle, Belfast, Manchester, Cardiff, Cambridge, Oxford, Southampton and Imperial College London. These regional centres provide physical resources and information for applications developers to develop grid based applications. A typical grid based application involves the use of information and resources at different locations. The distributed resources are integrated seamlessly. Large volumes of data can be transferred through the grid rapidly.

In the future, ordinary users and businesses will be able to tap into the grid to gain access to applications and services that require supercomputers, data warehouses, and physical appliances located somewhere in the world. They will gain access using the same Internet connections, such as a desktop computer, digital TV, or mobile phone.

9.1.6 Distributed Virtual Environments

Many business activities involve collaboration between people located in different places. Distributed Virtual Environments (DVEs) will be widely used in the future for remote collaboration and social interaction. Applications will range from business meetings and familiar gatherings to on-line teaching and learning, multi-participant games and collaborative design and production.

DVEs combine distributed communication and virtual reality technologies. They support real-time interaction between tens or hundreds of participants in the virtual world. A typical DVE is composed of many inter-connected computer servers. Each server is responsible for multiple clients who want to be part of the DVE. All servers receive constant updates from different clients about their current position and orientation, and then deliver this information to other clients in the virtual world. The servers, between them, also need to perform other tasks, such as object collision detection and synchronisation control. A large scale DVE system needs to support many clients and this imposes a heavy requirement on networking resources and computational resources. Therefore, there is usually a fundamental trade-off between scale, consistency and complexity.

DVE applications involve synchronous interaction between many users. To achieve high quality communication requires a sense of eye contact and a focus of attention between the participants as well as voice, expression and gestures. All these place great demands on the processor and network resources that can be radically different from passive broadcasting or asynchronous data applications.

Avatars are widely used in DVE applications. An avatar is a visual representation of a person in a virtual environment. Its appearance usually resembles a human. Some avatars even have behaviours in the virtual environment similar to people's behaviours in the real physical environment.

The construction process involves extensive interactions between people located in different places. DVE will provide ideal solutions to the communication problem.

9.1.7 Adaptive and Learning Systems

Ever since computers were invented, people have been fascinated by the prospect that computers can be intelligent, that they could solve problems, including new problems, by themselves with no human intervention. This area is broadly defined as Artificial Intelligence (AI). The American Association for Artificial Intelligence defines AI as 'the scientific understanding of the mechanisms underlying thought and intelligent behaviour and their embodiment in machines'. Until recently, due to the limitation of computer processing power, the number of AI applications has been limited. As the performance of computers continues to improve rapidly, computer systems will become more intelligent to such an extent that they will be able to learn and become adaptive. Technologies that will help to achieve this vision include fuzzy logic, neural networks and intelligent agents.

Fuzzy logic provides a unique method of approximate reasoning in an imperfect world. It differs from the conventional Boolean logic by allowing answers to a question that are not just right or wrong. The fuzzy principle states that everything is a matter of degree. For example, when people say that 'John is tall', they are expressing a judgment rather than a universal truth. In reality, there is no precise definition of 'tall' or 'short' for a person's height. It is usually a degree between two extreme. For example, most people would say a person with a height of 2 metres is 'tall' and a person with a height of 1.4 metres is 'short'. The vast majority of the population fall somewhere between these two extremes. A person with a height of 1.85 metres may be considered 'tall' by 80% of the people. It is difficult to express this fact using Boolean logic because the person is neither 'tall' nor 'short'. Fuzzy logic allows grey (not black and white) answers, by saying this person is 80% 'tall'. Most problems in real life are like this example, e.g., 'grass is green', 'today is hot', and 'the city is big'. Similarly, fuzzy logic will be very useful in solving the problems of the construction industry, for instance to judge 'how good a design is?' or 'how much a project costs?'

Neural networks refers to artificial neural networks, which is a problem solving technique that mimics people's brains. The human brain contains about 10 billion nerve cells, or neurons. Each neuron is connected to many other neurons. The brain's network of neurons forms a massively parallel information processing system. A complex problem is solved by the joint effort of many neurons working simultaneously. A neural network is a (much) simplified brain neuron structure and shares an important characteristic of the human brain – the ability to learn. Neural networks are good at recognising pattern from a collection of raw data. They are particularly suited to problems whose solution is complex and difficult to specify but for which there is an abundance of data from which the correct response can be learnt. At present, neural network technology is used in the control of microwave ovens and jet engines. In the future, it will be widely used to develop adaptive and self-learning systems.

Intelligent agent technology originates from Distributed Artificial Intelligence. An agent is an autonomous piece of software capable of controlling its own decision-making and acting, based on its perception of its environment, in pursuit of its user's objectives. Agents operate without the direct intervention of humans or others, and have some kind of control over their actions and internal state. Different agents can interact with each other and sometimes they can collaborate to achieve a common goal. Agents can be either reactive or proactive to their environments. Some agents can be mobile. They travel autonomously on a computer network. Agents can be divided into several types depending on the function they perform: collaborative agents, interface agents and information agents. Collaborative agents co-operate or negotiate with other agents or people. Interface agents are programs, like personal assistants, which learn about their users in order to perform tasks on their behalf. Information agents offer a potential solution to the 'information overload' problem – they find, collate and manipulate information from many distributed sources.

9.2 ACCELERATING CHANGE IN CONSTRUCTION

In the UK, the Strategic Forum for Construction chaired by Sir John Egan produced the 'Accelerating Change' report in 2002. Its aim was to promote the implementation of the Rethinking Construction principles. It set improvement targets for reduction of project capital cost, construction time, building defects, and accidents, and increases in predictability, productivity, turnover and profit. The report also outlines the instruments for achieving these targets (Figure 9.2)

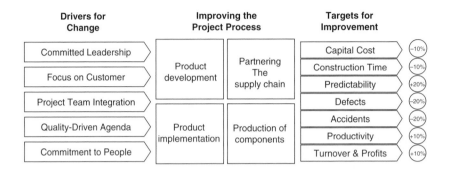

Figure 9.2 Rethinking Construction improvement targets

Partnering and integrated project teams are key to the success of the Accelerating Change initiative. The report set a strategic goal that 20% of construction projects by value should be undertaken by integrated teams and use supply chains by 2004. The figure should increase to 50% by 2007. IT and the Internet are viewed as key enablers to realise the necessary changes. The industry's desire to improve and its clients' demand for betters products and services, are 'business pull' for wider use of IT in the future in this sector.

9.3 VISION OF CONSTRUCTION IT

9.3.1 Questions for a Company Developing a Vision

People are very imaginative and can quickly see the possibilities for using technology. They do not always consider the factors necessary for getting it onto the market, or the problems that can arise. Some early inventions that took time to become successful including: the fax machine invented in the 1800s but only widely used since 1980, virtual reality – demonstrated in 1968 but only becoming a useful tool in the 1990s, and object models of buildings demonstrated in the 1970s but still not in general use.

A vision is only attainable if it is developed for a defined concept and if the reasons why it might not be achieved are identified. To develop a vision for the future a business might ask: What is its position in the market today, and where do we want to be in the future? Any company in construction should relate its own vision to that of a changing industry, for which visions have been set by Latham and Egan in the UK, and by numerous gurus worldwide.. If there is no clear vision there can be turbulence in an aimless organisation, short-term planning and infighting between those with their own ideas. The company vision needs to be shared between all those whose efforts are needed to make it work.

9.3.2 A Vision for the Construction Industry

A vision for the construction industry's future use of IT was developed by the IT Construction Best Practice group and Construct IT (2001). It has seven major themes:

- Model driven as opposed to document driven information management on projects
- Life cycle thinking and seamless transition of information and processes between phases
- Use of past knowledge in new developments
- Dramatic changes in procurement philosophies as a result of the Internet
- Improved communications during all life cycle phases via visualisation.
- Increased opportunities for simulation and 'what if' analysis
- Increased capabilities for change management and process improvement.

Each company setting itself such a vision needs to relate these areas to its own business and to the phases of the construction process in which they are involved. Wide consultation is necessary with staff, business partners and customers, and a clear plan needs to be established with frequent reviews as new solutions become available.

9.3.3 Barriers to Success

A study of the reasons why visioning can fail (Contrada, 1998) identifies the following statistics:

- Only 5% of the workforce understands the strategy
- Only 25% of managers have incentives linked to strategy
- 60% of organisations do not link budgets to strategy
- 85% of executives spend less than one hour per month discussing strategy
- 90% of companies fail to execute the strategy.

Information technology is only a part of such strategies but, as one of the fastest changing aspects of modern business, it can be used as a catalyst for change. People are often resistant to change and are usually the reason for systems failing or taking too long to be adopted, hence the need for wide consultation.

9.3.4 Stages in the Development of a Vision

The Construct IT report 'How to develop a strategy plan' (Construct IT, 2000) suggests dividing the process into five stages:

- Initiate the strategy planning process
- Identify the business position
- Examine the systems and technologies available on the market
- Develop a system and technology roadmap
- Prioritise solutions.

These can be further subdivided and responsibilities given to individuals in order to ensure success. Experts from outside a company can be invited to give their views but conflicting commercial interests should be balanced. In a research project at the Technical University of Denmark (Howard, 2001) future visions were needed for the technological context in which a new building classification system would be used. Groups of experts were asked for their visions of the next 5 and 10 years in relevant fields. The possibilities were relatively easy to state but they found more difficulty with the conditions necessary for success and the barriers that might prevent the technologies being taken up. The views of these experts were then tested against opinions from a more typical selection of firms from the construction industry that would be needed to buy into new technologies when they became available.

From all the ideas put forward in this book it should be possible for an individual or a company to develop a vision of how technology could serve them in whatever part of construction they work. This is probably the easiest part of the task. Getting colleagues to agree and share the vision is rather harder, revising it to keep up with rapid technology change can be another problem, and achieving the whole vision is well nigh impossible, but it must be attempted if the construction industry is to become more productive and deliver better buildings.

9.4 EMERGING TRENDS

As computing technology continues to develop, the following trends begin to emerge:

- Construction companies recognise the value and importance of knowledge assets. Knowledge management systems will be more widely used at the project, organisation and industry levels.
- Object-oriented CAD systems will become the norm. They will be used as a design and modelling tool, rather than just a presentation tool.
- Physical performance simulations will become part of virtual prototyping during the design development process.
- E-tendering and e-procurement will become widely used.
- 4D (3D plus time dimension) modelling will provide a visual tool for project planning and scheduling. Mobile devices will be widely used at construction sites.

- More intelligent devices will be used to control the operation and maintenance of buildings in an integrated fashion.
- Product data technology will become mature. Data and process standards will eventually lead to whole life cycle integration.

In addition, IT will have an impact on the organisational structure of the construction industry and people's working patterns:

- IT will help to break down the traditional barriers between different organisations. It will facilitate the sharing of knowledge and resources between partners in a supply chain. This will leads to new forms of Virtual Organisation.
- The Intranet used by a business may become the defining factor in who belongs to the company, rather than any particular location. Location has become unimportant when ordering goods over the Internet or by telephone today. The enquiry will probably be handled at a Call Centre where many operators are based and serve many businesses. When goods are dispatched they will leave an automated warehouse, probably shared by different firms, and the administration of the business could well be from someone's home or based in another country.
- Project Extranet or Project Web will allow a virtual project team to share project information, but it allows the teams to meet less frequently or to be based in different locations, and to call quickly on specialist advice. An extension of this is the subcontracting by information businesses to other time zones or places where labour is cheaper.
- Project Extranets also give project team members 24/7 (24 hours a day, 7 days a week) access to project information. Tasks can be carried out anywhere and any time, not just during working hours at the offices.

9.5 ON-LINE RESOURCES

http://www.technology-watch.info/
Technology Watch provides regular updates on a wide range of new and emerging information and related technologies relevant to the construction industry. These technologies include those that have immediate application, as well as 'blue sky' technologies that are still in development but have future potential.

http://www.computer.org/pervasive/
IEEE Pervasive Computing, freely available from this site, delivers the latest peer-reviewed developments in pervasive, mobile, and ubiquitous computing to developers, researchers, and educators who want to keep abreast of rapid technology change.

http://www.personal-ubicomp.com
Personal and Ubiquitous Computing has published some of the most innovative international research contributions on the design and evaluation of new generations of handheld and mobile information appliances.

http://www.redwoodhouse.com/wearable/
This site provides many useful links to wearable computers technology and products.

http://www.nextwave.org.uk
Next Wave Technologies and Markets is a UK DTI sponsored programme. It aims to ensure that UK business is structured and equipped to exploit new information and communications technologies and products that enable intelligent functionality to be embedded into devices that will eventually become an integral part of daily life.

http://www.media.mit.edu/
The media laboratory at MIT conducts research on the impact of new IT and communication technology on the built environment and aspects of people's interaction with the environment.

http://www.equator.ac.uk/
Equator is a six-year Interdisciplinary Research Collaboration (IRC) supported by EPSRC that focuses on the integration of physical and digital interaction.

http://cooltown.hp.com/
The HP Cooltown Project investigates the intersection of nomadicity, appliances, networking, and the web. It presents a vision of a technology future where people, places, and things are first class citizens of the connected world, wired and wireless – a place where e-services meet the physical world, where humans are mobile, devices and services are federated and context-aware, and everything has a web presence.

http://research.microsoft.com/easyliving/
Microsoft EasyLiving is developing a prototype architecture and technologies for building intelligent environments.

ftp://ftp.cordis.lu/pub/ist/docs/istagscenarios2010.pdf
An on-line report on Ambient Intelligence Scenarios from the EU. This report starts with four scenarios that illustrate how Ambient Intelligence might be experienced in daily life and work around 2010. The concept of Ambient Intelligence (AmI) provides a wide-ranging vision on how the Information Society will develop. The emphasis of AmI is on greater user-friendliness, more efficient services support, user-empowerment, and support for human interactions.

http://www.bluetooth.com
This is the official Bluetooth website. Bluetooth is a wireless technology that enables different electronic devices to communicate with each other. It will help to create a more intelligent living environment.

http://www.mobileinfo.com
This site contains technical information on mobile computing and information on commercial products.

CONCLUSION

The continuing advance of IT and the ever more pervasive use of computers will transport the traditional construction industry into the information age. Soon, e-business and collaborative working systems will alter the interaction between supply chain partners. Mobile computing will give on-site workers access to updated project information. Integration technology will improve the information and knowledge sharing throughout the whole life cycle of construction projects. Intelligent and ambient sensors will make smart buildings a reality. Harnessing the potentials of IT with the needs of the industry and its clients will bring positive changes to, not only the construction process, but also the product of the process – the built environment itself.

DISCUSSION QUESTIONS

1. What are the benefits of increased network bandwidth for construction?
2. How can distributed virtual environments be applied to the design and construction process?
3. Give some ideas about which tasks in construction could be aided by adaptive and learning systems.
4. Develop a vision for your own use of IT in your work.

REFERENCES

Akintoye A. and Fitzgerald E., 2000, A survey of current cost estimating practices in the UK, Construction Management and Economics (2000) 18, 161-172

Ali K.N., Sun M., Petley G. and Barrett P., 2002, Improving the business process of reactive maintenance projects, Facilities, Vol. 20, No. 7/8, 2002, pp251-261

Architectural Record, 1985, The results of RECORD survey: How are firms with computers faring – and what are the nonusers waiting for?, Architectural Record, June 1985, pp37-41

Amor R. and Anumba C., 1999, A survey and analysis of integrated project databases, Proceedings of the Second International Conference on Concurrent Engineering in Construction, Espoo, Finland, 1999, pp217-228

Armstrong J., 2002, Facilities management manuals: a best practice guide, CIRIA publication C581, London 2002

Augenbroe G. and Laret L., 1988, COMBINE pilot study report, Research paper

Baldwin A., Betts M., Blundell D., et al, 1999, Measuring the benefits of IT innovation, in Betts (eds), Strategic management of IT in construction, Blackwell Science, 1999

Bazjanac V., 1975, The promises and the disappointments of Computer Aided Design, in Nicholas Negroponte, editor, Computer Aids to Design and Architecture, Mason/Charter, London

Becker F., 1990, The Total Workplace, Van Nostrand Reinhold, New York, NY

Bentley R., Appelt W., Busbach U., et al, 1997, Basic Support for Cooperative Work on the World Wide Web, International Journal of Human-Computer Studies 46(6): Special issue on Innovative Applications of the World Wide Web, 827-846, June 1997

Betts M., 1999, Strategic management of IT, Blackwell Science

Björk B., 1987, The integrated use of computer in construction – the Finish experience, Conference proceeding ARECDAO87, pp18-21

Brambley M.R., Addison M.S., et al, 1988, Advanced energy design and operation technologies research, Pacific Northwest Laboratory research proposal, PNL-6255, UC-95d

Bryan M., 1997, An Introduction to the Extensible Markup Language (XML), The SGML Centre, 1997, http://www.personal.u-net.com/~sgml/xmlintro.htm

Building Centre Trust, 1990, Building IT 2000, Report

Building Centre Trust, 1999, IT usage in the construction team, published by Construction Research Communications Ltd, London

Bush I., 1999, A Proposal for a Feasibility Study for an Application of Knowledge Management, MSc coursework report, School of Construction and Property Management, University of Salford, UK

Capron H.L., 2000, Computers: tools for an information age (6th Edition), Upper Saddle River, N.J.: Addison-Wesley

Chartered Institute of Building (CIOB), 1982, Maintenance Management – A Guide of Good Practice, CIOB Publication

CICA, 2003, Software Directory, Construction Industry Computing Association www.cica.org.uk

CICA, 1993, Lasted industry survey of leading firms, Building on IT for quality, CICA report

CICA, 1996, Computing for Site Managers, CICA report

Clark G. and Mehta P., 1997, Artificial intelligence and networking in integrated building management system, Automation in Construction, 6 (1997) pp481-498

Construction Computing, Issue No. 60, February 1997

Contrada M.G., 1998, Maximising corporate performance with the balanced scorecard. Renaissance: The Balanced Scorecard Collaborative, USA

Davenport T.H., 1993, Process Innovation – Reengineering Work through Information Technology, Harvard Business School Press

Department of Environment, 1995, Construct IT Bridging the Gap, An information technology strategy for the United Kingdom construction industry, Department of Environment, UK

Department of the Environment, Transport, Region (DETR), 2000, Construction Statistic Annual 2000 Edition

Eastman C.M., 1999, Building product models, CRC press, USA

Egan J., 1998, Rethinking Construction, HMSO

Fernando T., 1997, Virtual Environment, MSc IT Management in Construction Module Workbook, University of Salford, UK

Finch E., 2000, Third-wave Internet in facilities management, Facilities, Vol 18, No 5/6, pp204-212

Gartner Group, 2001, e-business chart by the Gartner Group reported in Ingeniøren.

Greenberg D.P., 1985, Computer graphics and visualisation, in A. Pipes (edited), Computer-Aided Architectural Design Futures, Butterworths

Howard R., et al, 2001, Research on international experience and future developments in IT, Building classification centre contract. BYG.DTU for Teknologisk Institut

Howard R., 1996, Computing in construction – pioneers and the future. Butterworth Heinemann

IAI, 1999, Introduction to IAI, IAI CD release version 2.0, March 1999

ISO, 1994, ISO 10303-11, EXPRESS language reference manual, International Standardisation Organisation, Geneva, Switzerland, 1994

Kharrufa S., Aldabbagh H. and Mahmond W., 1988, Developing CAD techniques for preliminary architectural design, Computer Aided Design, Vol. 20, No. 10, 1988, pp581-588

Latham M., 1994, Constructing the team, HMSO

Laudon K. and Laudon J., 2000, Management Information Systems, Prentice Hall International, www.prenhall.com/laudon

Le Corbusier, 1927, Vers une architecture. Architectural Press, London

Mitchell W.J., 1995, City of bits space, place, and the Infobahn, MIT Press

Negroponte N., 1995, Being digital, Hodder & Stoughton

NHS Estates, 1996, Re-engineering the facilities management service, Health Facilities Note 16

Nonaka I. and Takeuchi H., 1995, The Knowledge-Creating Company: How Japanese Companies Cretae the Dynamics of Innovation, NewYork; Oxford: Oxford University Press

Nonaka I., Reinmoeller P. and Senoo D., 2000, Integrated IT Systems to Capitalize on Market Knowledge, in von Krogh G., Nonaka I. and Nishiguchi T. (ed.), Knowledge Creation – A Source of Value, London: Macmillan Press Ltd

Paulson B.C., 1995, Computer applications in construction, McGraw Hill Inc

Porter M., 1980, Competitive strategy, Free Press, New York

Reynolds R.A., 1993, Computing for Architects (2nd Edition), Butterworth-Heinemann Ltd

RIBA Journal, January 1995

RIBA, 2000, Architect's plan of work for the procurement of feasibility studies, RIBA publications, London

Richens P., 1990, MicroCAD software evaluated for construction, Published by Construction Industry Computing Association, Cambridge

RICS, 1988, SMM7: Standard method of measurement of building works, Royal Institution of Chartered Surveyors

Seeley I., 1976, Building Maintenance, MacMillan Press Ltd, Basingstoke

Smith A.J., 1995, Estimating, tendering and biding for construction, Macmillan Press Ltd, London

So A.T.P., Chan W.L. and Tse W.L., 1997, Building automation system on the Internet, Facilities, Vol. 15, No. 5/6, pp125-133

Spedding A., 1994, CIOB Handbook of Facilities Management, Longman Scientific & Technical, England

Spon, 2001, Spon's architects' and builders' price book, Spon, London

Stevens G., 1991, The impacts of computing on architecture, Building and Environment, Vol. 26, No. 1, 1991, pp3-11

Sun M. and Lockley S.R., 1997, Data exchange system for an integrated building design system, Automation in Construction, No. 6, 1997, pp147-155

Sun M., Aouad G., Bakis N., et al, 2000, GALLICON – A prototype for the design of water treatment plants using an integrated project database, International Journal of Computer-integrated Design and Construction, Vol. 2, No. 3, 2000, pp157-165

Tay L. and Ooi J.T.L., 2001, Facilities management: a "Jack of all trades?", Facilities, Vol. 19, No. 10, 2001, pp357-362

Then, D.S.S., 1999, An integrated resource management view of facilities management, Facilities, Vol. 17, No. 12/13, pp462-469

Turban, McLean and Wetherbe, 2002, Information technology for management, Wiley

Wang S.W. and Xie J.L., 2002, Integration building management system and facilities management on the Internet, Automation in Construction 11(2002) pp707-715

Warthen B., 1989, A history of AEC in IGES/PDES/STEP, Computer Integrated Construction, Vol. 1, Issue 1, 1989, pp4-8

INDEX

3D models
  Solid model, 63
  Surface model, 63
  Wireframe, 62
3G, 168

Abacus, 3
Accelerating Change, 173, 174
Adaptive system, 172
ADSL, 10, 140, 167, 168
AEC/FM, 154, 164
AI, 172
Altair 8800, 4
Animation, 18, 64, 67, 74
Annotations, 58
ARPANET, 4, 9
Artificial intelligence, 172, 173

Barbour Index, 44, 109
BCIS, 106, 109
Benchmarking, 28
Bill of Quantities, 103
Blaise Pascal, 3
Bluetooth, 145, 168, 178
BoQ, 103
BREEAM, 80
British Standard, 59
BS 1192, 59
Building engineering applications,
  16, 18
Building Management System, 86
Building operation management,
  137
Building Regulations, 82
Building simulation, 80
Build-Own-Operate-Transfer, 28
Business Process Reengineering, 25,
  46, 149

Cable modem, 167

CAD, 15, 18, 19, 20, 51, 53, 54, 56,
  58, 60, 68, 74, 78, 102, 103, 104,
  127, 134, 137, 138, 147, 149,
  153, 154, 156, 159, 160, 161, 166
  2D, 53
  3D, 18, 54, 55, 62, 64, 66, 68, 73,
    74, 75, 78, 86, 134, 160
CAD link, 104
CAD object
  Layering, 58
CAD objects
  arc, 57
  Blocks, 58
  Ellipse, 57
  External refernces, 58
  Polyline, 57
CADLink, 19, 89, 104, 109
CAFM, 134, 138
Case Based Reasoning, 36
CBR, 36
CD-ROM, 4
Charles Babbage, 3
CIBSE Guide, 81, 144
CIE Overcast Sky, 84
Coaxial cable, 167
Collaborative working system, 40
Communication, 2, 126, 139, 162
  Wireless, 126
Computer Aided Design, 16, 18, 51,
  53, 180
Computer Network, 6
  Capacity comparison, 168
  Computing grid, 170
  Extranet, 10, 177
  Internet, 2, 3, 5, 7, 9, 10, 27, 38,
    40, 41, 43, 46, 47, 60, 127,
    134, 136, 138, 140, 160, 161,
    162, 175, 177
  Intranet, 10, 177
  Local Area Network, 7
  WAN, 7
  Wide Area Network, 7

Computer Technology, 3
Computers
 Benefits of, 5
 Generation of, 3
 Hand-Held, 127
 Network. *See* Computer Network
 Pen Tablet, 127
 Timeline of computing, 4
 Touch Computers, 127
Construction industry
 Computer ownership, 14
 Computerisation, 13
 Features of, 11
 IT applications, 16
 IT usage, 15
 IT vision, 175
 Management techniques, 28
Construction process, 1, 11, 17, 28,
 90, 126, 153, 157, 175
 IT roadmap, 17
Construction Site Monitor, 20, 125
Coordinate system, 54, 55
 Absolute coordinate, 55
 Polar coordinate, 55
Cost analysis, 107
Cost estimating, 19, 92, 93, 95, 96,
 100, 102, 103, 104, 158, 159,
 160, 161
 Equipment cost, 94
 Indirect cost, 95
 Labour cost, 94
 Unit Cost Estimating, 93
 Unit rate method, 93
Critical Path Method, 113, 115
CRM, 29
CSCW, 40
Cymap, 19, 84, 86

Data, 26
Data standards, 21, 154
Daylighting, 83, 84, 91
Desk Top Publishing, 21
Digital age, 3
Digitiser, 104

Distributed Virtual Environment,
 172
Drawing template, 56
DVE, 172
Dynamic Data Exchange, 102

Earned value table, 123
E-business, 3, 41, 44, 46
 Growth of, 41
EDM, 17, 39, 40
Egan, 28
Electronic Document Management,
 17, 39
E-mail, 2, 10, 15, 21, 41, 127
Energy analysis, 81
ERP, 29
E-science, 171
Esti-mate, 102, 110
EVEREST, 102
EXPRESS, 152

Facilities Management, 16, 20, 132,
 133
 Asset management, 136
 BIFM, 133
 Definitions, 133
 Future trends, 143
 IFMA, 133
 Space management, 136
Fibre optic, 167
FM, 20, 132, 133, 134, 135, 136,
 137, 138, 142, 147
Formula, 96
FTP, 10
Fuzzy logic, 172

GALLICON, 158, 159, 160, 161,
 162, 163, 182
Gantt charts, 115
GIS, 76, 130, 134
Gottfried Wilhem von Leibniz, 3
Graphic Information System, 134
Grid, 58
Griffiths, 106

Head-Mounted Display, 69
Helpdesk, 135
  On-line system, 138
Hevacomp, 19, 82, 85, 86
HVAC, 19, 81, 85, 158

IAI, 154, 155, 164, 181
IBM PC, 4
IFC, 21
Information, 26
Information Age, 1
Information management, 25, 38
Integrated Circuit, 3
Integrated Project Database, 157
Integration, 21, 29, 148, 150, 152,
  157
  Benefits of, 162
  Methods of, 150
  Modelling, 152
  Need for, 149
  Process models, 153
  Product models, 152
Intelligent Buildings, 144
Internet Service Providers, 10
Isambard Kingdom Brunel, 80
ISDN, 10, 140, 167, 168
ISP, 10
IT Best Practice Programme. *See*
IT management, 45

KBS, 35
Knowledge, 26, 34
Knowledge conversion
  Combination, 34
  Explicit knowledge, 33
  Externalisation, 34
  Internalisation, 34
  Tacit knowledge, 33
Knowledge management, 34
  Knowledge conversion, 33

LAN, 7, 9, 10, 126
Latham, 28
Laxton, 106

Le Corbusier, 79
Lighting analysis, 81, 83
Lotus, 96

Management theory, 26
MasterBill, 102
Material procurement, 125
Microsoft Excel, 29, 96, 97
Microsoft Windows, 5
MIS, 27
Mobile Broadband, 168
Mobile network, 167
Modem, 5
Moore's Law, 166
MS-DOS, 4

Network topology
  Linear Bus topology, 7
  Ring topology, 7
  Start topology, 7
Neural network, 173

Object Linking and Embedding,
  102, 148
OCR, 104
Office management, 28
On-line services, 2

Partnering, 28, 32
PDAs, 168
Personal Computers, 4, 96
Personal Digital Assistant, 5, 168
PERT charts, 113
Pervasive computing, 169
Planning, 19
Portals, 43
Porter's model, 30
Price Book, 106
Programming languages, 3
  BASIC, 4
  COBOL, 3
  FORTRAN, 3
Project planning, 112
Public-private partnerships, 28

Quantity take-off, 19

RAM, 4, 5
Reandering
  Scanline, 66
Rendering, 65
  Radiosity, 66
  Ray tracing, 66
  Texture mapping, 66
RIBA Plan of Work, 11
RIPAC, 102, 103, 109

Scheduling, 19, 112
  Milestone, 118
  Recurring task, 118
  Task dependencies, 118
  Task hierarchy, 117
SDSL, 167
Site management, 1, 19, 124
Site operation, 20
SMM7, 93
Snap, 58
Speech recognition, 71, 169
SPON, 94, 106
Spreadsheet, 96
  Charts, 99
  Evaluation, 100
Spreadsheet Applications
  Customised applications, 100
Standard Assessment Procedure, 85
STEP, 21, 154, 155, 156
Structural Analysis, 86
Supply chain management, 28

TCP/IP, 9, 10
Telepresence, 170, 171
Teleworking, 3

Ubiquitous Computing, 169
USENET, 10
U-value, 81

VBA, 100
Virtual Private Network, 11, 140
Virtual Reality, 18, 68, 75, 160
  BOOM, 70
  Camparison with animation, 74
  CAVE, 72
  Comparision with CAD, 74
  Data gloves, 73
  Fully immersive *VR*, 69
  Input devices, 73
  Non-immersive VR, 69
  Reality Centre, 71
  Semi-immersive VR, 69
  Workbench, 71
Visualisation, 1, 16, 51, 54, 64, 67, 74, 78, 85, 90, 152, 159, 160
VR, 18, 68, 69, 71, 74, 75, 152
VRML, 73, 159

Wessex, 106, 108, 110
Work Package Structure, 113
World Wide Web, 2, 5, 10, 38

Y-value, 81